THE
MX ICBM
AND
NATIONAL
SECURITY

THE
MX ICBM
AND
NATIONAL
SECURITY

Colin S. Gray

PRAEGER

PRAEGER SPECIAL STUDIES • PRAEGER SCIENTIFIC

Library of Congress Cataloging in Publication Data

Gray, Colin S.
    The MX ICBM and national security.

    Bibliography: p.
    Includes index.
    1. MX (Weapons system)   2. United States—National
security.   I. Title.
UG1312.I2G72   358′.174′0973   81-2557
ISBN 0-03-059442-1   AACR2

Published in 1981 by Praeger Publishers
CBS Educational and Professional Publishing
A Division of CBS, Inc.
521 Fifth Avenue, New York, New York 10175 U.S.A.

© 1981 by Praeger Publishers

23456789   145   98765432

Printed in the United States of America

PREFACE

This book is intended as a contribution to, not a leisurely commentary upon, a major defense debate. The price that must be paid for policy relevance is the acceptance of a degree of technical and political dynamism in the real world that may render individual sentences and even perhaps whole paragraphs in this book dated by the time of publication. The problem can be minimized—it cannot be eliminated altogether. Public policy controversy will not remain conveniently (for the author) quiescent during the period of lead time for book production.

As of the time of writing, the author has, as fixed points of reference, a mature MX ICBM program and basing scheme—the final Carter design—and a clear understanding of the principal concerns that the Reagan administration has regarding that inherited program. One cannot predict with confidence just which technical changes, in detail, the Reagan administration will decide to effect. Moreover, notwithstanding the passion that has been, and no doubt will continue to be, aroused among missile engineers (professional and amateur) over the design of "the best" MX/MPS system, this author is convinced that such passion over essentially secondary matters, however inevitable and perhaps even desirable (technical details are important), is relatively unimportant for the purpose of this book.

Some long-standing friends of the author (for example, William R. Van Cleave and Paul H. Nitze) believe intensely in vertical, as opposed to horizontal, sheltering for MX ICBMs—to isolate but one technical issue (though one with many ramifications). The author prefers, on balance, horizontal sheltering, but he certainly believes that vertical shelters would be "good enough."

This is neither a work of history nor a work of journalism. The historian forfeits the opportunity to influence events, while the journalist is so concerned to be accurate on a day-to-day basis that he or she may easily fall prey to fashion and neglect the shape of the woods for the joy of placing each individual, newly discovered tree. This is a work of policy analysis written by a working strategist— really by a participant-observer. The author has close friends on all sides of the MX/MPS debate (he hopes they remain close friends after this book is published). He has followed the dramatically shifting fortunes of MX/MPS from its inception and has endured more basing-scheme changes than he cares to recall.

To return briefly to the theme of passion cited earlier, a measure of anger underlies this book. It was obvious to the author, and to many of his friends, as early as 1974-75 that a survivably based MX ICBM was needed with great urgency. Contemporary with the publication of this book is the fact of the nearly total vulnerability of the U.S. silo-housed ICBM force. This should never have been allowed to happen. It was predictable, and it was predicted. There is no little degree of frustration in being compelled to acknowledge that anybody's MX/MPS scheme, as of 1981, is not properly responsive to Soviet silo-threatening capability, when one has been urging expedition in MX/MPS development and deployment for more than six years.

Many people in government, the armed services, and academic life have contributed to my thinking on the subject of this book; they know who they are and do not need individual mention here. However, individual mention is needed for my assistant, Audrey Tobin, for the "cruel and unusual punishment" of coping with my text and with the index, and for my wife, Valerie, for her surprisingly cheerful toil on the bibliography. Alas, there is no evading the standard statement to the effect that the author alone is responsible for what is good and bad in this book.

CONTENTS

INTRODUCTION

The MX ICBM system has been endorsed at the highest level of the U.S. government. President Carter endorsed full-scale development of the MX missile in June 1979, and he approved the horizontal-dash basing mode in September of that year. These policy facts have not, however, stilled the debate over either the defense necessity for the MX ICBM system, or the details of the basing mode and its desirable operating characteristics.

MX is proving to be unusually controversial as a weapon-system, though the controversy covers so wide a range of public concerns, and embraces such divergent though still largely astrategic assumptions,[1] that the debaters seem more often to be talking past, rather than at, each other. In fairness, there appears to have been a lack of clarity on both sides of the MX debate—and, perhaps above all else, a lack of clarity on each side over the real concerns of the other. For the major example of this phenomenon, it can be difficult to secure a favorable reception to the proposition that horizontal-dash MX is the best available solution to the land-based missile vulnerability problem if the audience (a) does not believe that there is a problem in need of solution, or (b) acknowledges the problem (at the strategic-technical level) but deems the finding of a solution a matter of little importance or urgency.

The purpose of this study is to address objectively all of the more important political-strategic questions that have been, or should be, asked of the official proponents of the MX ICBM system. The author freely admits to being strongly in favor of the system. However, that attitude flows from analysis conducted over many years and does not, in a real sense, detract from the objectivity of this study. Objective analysis carries no implications for the conclusions reached. An inconclusive paper is not, ipso facto, more objective than a paper that states a clear, preferred policy outcome.

As will be demonstrated, many of the questions raised concerning MX are neither trivial nor reflective of unpatriotic motives. It may be safely assumed that both sides of the debate are genuinely dedicated to the promotion of the national security of the United States. Unfortunately, once one proceeds beyond the "motherhood" kind of affirmations, one discovers differences in attitudes toward the nature of world politics, the probable or possible character of international relations in the 1980s, and the requirements of deterrence (at a fairly deep philosophical level), such that the U.S. defense-policy debating community almost certainly will not be

ix

able to achieve a community-wide consensus on the wisdom of MX deployment. Nevertheless, analysis such as that presented in this study may help to force a robust working majority sufficient to see the MX project through—a majority whose belief in the MX program would be based on a good enough appreciation of the political and strategic issues pertinent to the program so that program difficulties "down the road," unless they were of a truly major character, would be unlikely to erode the MX consitituency very markedly.

A pragmatist might argue that all that matters is that the MX program proceed—no matter what the reasons for the endorsement. He or she might continue with the plausible, indeed accurate, claim that as a weapon program proceeds and gathers momentum, its "interested" constituency grows—the number of people and organizations with a nonmarginal stake in the program expands. However, although major defense program debates are rarely won or lost on the basis of the merits of contending strategic arguments,[2] the absence of a solid strategic case in which one can believe constitutes a fundamental weakness in a program. Such absence may not mean that there is no solid strategic story to be told; rather, it may reflect nothing more than the fact that the weapon-sponsoring coalition did not take the trouble to think through the political-strategic rationales for its program. The easier the acquisition of official and congressional assent to a program, the stronger the temptation not to "look for trouble" by unnecessarily raising strategic policy questions that—though supportive of the program—may nourish some germ of policy-level criticism or skepticism that could have been avoided.

It is important that MX proponents not elect to sit back content on the strategic policy front, enjoying the glow of White House and congressional approval. Favorable though the domestic political climate is for MX (with some reservations of the "not in my backyard" variety), courtesy of Soviet misbehavior that climate could change. Moreover, if a president does the right thing (like endorsing MX), but for the wrong reasons (like preemptively appeasing defense-minded senators who could vote either way on SALT II, seeking to impress Mr. Brezhnev, immediately prior to the Geneva summit in June 1979, with the quality of U.S. arms-competitive determination, or attempting to undercut the defense arguments of a rival presidential candidate),[3] his commitment to the system may be no more than seasonal.[3] MX proponents, though currently "victorious," should never forget that their system, according to one stream of logic, constitutes a direct affront to the still potentially important segment of the U.S. public-policy-influencing community that views silo-threatening capability as destabilizing.

The arms control sector of the U.S. national security community is currently "bunkered down," weathering the domestic

passions stirred by the Soviet invasion of Afghanistan[4] and antici-
pating a lengthy period "out in the cold" under the Reagan administration, but the arms control ideology, and beliefs in the necessity of
detente, have taken so firm a root since the 1960s that the MX program could be ambushed in the future.

There will be people immune to many of the arguments presented in this book—the alleged evils of deceptive ICBM basis, hard-
target kill-capable MIRVs, and new moves in the arms race are
revealed truths to a hard core of the arms control faithful. Nonetheless, those faithful are few in number, and probably becoming fewer.
Their arguments are addressed directly and seriously in this study,
not in the hope that they will see light on the road to Damascus, but
because the majority of the public attentive to defense issues, and to
MX in particular, is not equipped to answer them unaided. Save in the
immediate area of proposed MX deployment (presumably Nevada-
Utah), the general public is not a problem for the MX program. That
public, at the present time, is strongly in favor of the strengthening
of the U.S. defense posture; is (and has always been) relatively
ignorant of, and uninterested in, the details of that posture; and needs
no persuasion that the Soviet Union means us, our friends, and our
allies ill. The problem, or opportunity, lies among opinion leaders
at all levels, even including some defense professionals, who either
have not made up their minds about the wisdom of MX or have rejected or endorsed MX on somewhat casual or even transitory or highly
contingent grounds. For people such as these, the serious political-
strategic issues relevant to the MX ICBM program have to be addressed.

THE
MX ICBM
AND
NATIONAL
SECURITY

PART I
THE SETTING AND TERMS
OF THE MX DEBATE

1

THE POLITICAL
AND STRATEGIC CONTEXTS:
A ROUGH WORLD

Debate over a particular weapon system naturally features, on
the surface, such issues as cost effectiveness, timeliness in regard
to the threat, strategic missions, arms control impact, and the like.
However, that being said, each weapon debate is specific to time
and national mood—and, as a consequence, to the defense budgetary
circumstances specific to that time and mood. Several examples
from the 1970s illustrate this thesis. In 1968-72, ABM technology
was taken by radical and even moderate liberal lobbying groups to be
symbolic of a U.S. military-industry complex (MIC) that was alleged
to be running out of control. Negative domestic political fallout from
Vietnam poisoned the well of defense debate. Sentinel and then
Safeguard could not be debated fairly.[1]

The B-1 bomber decision of June 1977 is another example of
political circumstance dominating rational defense considerations.
Mr. Carter, newly elected and pressing hard for "real arms control," for nuclear nonproliferation, and for human rights (all worthy
and important goals, but all really secondary to the need to keep
Soviet tanks out of additional territory around the Eurasian rimland),
did not bring himself to deny his own well-known feelings about nuclear
weaponry and affront constituencies who expected a new moral impera-
tive in U.S. foreign policy. If the B-1 bomber had come up for a
production decision in 1979 or 1980, rather than 1977, there can be
little doubt that the president would have approved it. Fortunately
for Mr. Carter, he had the promise—though not the actuality—of the
cruise missile as a plausible alibi for his decision. Even so, the
case for the B-1 in 1977 (which, if anything, is even stronger in
1981) was embarrassingly persuasive.[2] Naturally, as with every
major weapon system, the B-1 was challengeable on some technical
grounds (particularly those pertaining to the cost effectiveness of its

supersonic dash capability). The B-1 episode illustrates two basic truths. First, political circumstance is at least as important as "rational" strategic analysis; and, second, one can always design a better weapon system tomorrow (that is, the quest for weapon perfection, the endless tinkering with a design, results in the absence of actual combat capability). As a consequence of the 1977 B-1 decision, the U.S. Air Force (USAF) today is nearly desperate for an enhancement of its manned penetrating bomber force for the mid-1980s.[3]

Mr. Carter's foreign and defense policy from 1977 until late 1979, with respect to its fundamental architecture, was simply wrong. This claim is noncontroversial and was admitted, very close to explicitly, both by the president himself and by his senior advisers. Mr. Carter, to his credit (for honesty), admitted that the Soviet invasion of Afghanistan accomplished more for his education concerning things Soviet than had the events of the previous (nearly) three years. Also, in a Department of Defense report, one finds the following statements (admissions?):

> Critical turning points in the histories of nations are difficult to recognize at the time. Usually they become clear only in retrospect. Nonetheless, the United States may well be at such a turning point today. We face a decision that we have been deferring for too long; we can defer it no longer. We must decide now whether we intend to remain the strongest nation in the world. The alternative is to let ourselves slip into inferiority, into a position of weakness in a harsh world where principles unsupported by power are victimized, and to become a nation with more of a past than a future.[4] (Emphasis added.)

Quite so! This statement by Harold Brown was a fundamental challenge to the corpus of opinion that held that the arms race itself was more of a problem for U.S. security than were Soviet intentions or actions. Scholarly research during the 1970s showed that each superpower has tended to produce and deploy the strategic weapons that it chooses on schedules largely unrelated to specific events abroad.[5] The lead time for research, development, test, evaluation, and production—the better part of ten years—is far too long, in both countries, for an action-reaction mechanism to function save only in macroscopic ways. The strategic missiles being deployed by the U.S.S.R. in 1981 were born as the era of strategic parity dawned in 1969-70. In short, the idea that there is a malevolent "arms race" that is the danger is almost entirely misleading. This is not to deny either that U.S. arms competitive actions may not (eventually) spur Soviet counteraction or that the arms race, per se, may not be fueled by such U.S. action. But it is to deny that the dominant "action-reaction" arms race model deployed for debating purposes in the late 1960s had any substantial basis in fact.[6]

The U.S. defense community as a whole agrees today that the Soviet Union endorses a war-waging/war-winning strategic doctrine;[7] acknowledges that even the concept of strategic parity poses grave problems for the credibility of U.S. extended deterrence commitments; and sees little or no hope for the "education" of Soviet policymakers in the Western ideas of a strategic stability dependent upon the maintenance of the vulnerability of Soviet and American societies. In short, unlike 1970, with the debate over Safeguard ABM and the concern over the MIRVing of U.S. ICBMs and SLBMs, few voices are raised against MX in the context of the arms competition (that is, how its deployment might fuel the competition). If anything, perhaps strange to say, the Department of Defense appears to be almost unduly relaxed on the subject of the possible arms-race consequences of the MX program. The truth of the matter is not that officials are embarrassed by their own logic—rather, it appears that they have not taken the time to consider the question.

If the United States proceeds to deploy MX in a survivable basing mode, what should the United States expect the U.S.S.R. to do about it? There is no conspiracy of official silence on this question; there are simply the facts of lead time and necessary ignorance. MX ICBM deployment is not imminent, and Soviet strategic competitive "style" remains largely terra incognita. Liberal-minded arms control critics of the MX ICBM program, such as William Kincade, are entirely right to raise the question of the probable character of the Soviet strategic response.[8] The absence of satisfactory answers profferred, to date, in response to Mr. Kincade's question ("Will MX Backfire?") appears to speak more to the unwillingness of the bureaucracy to risk providing additional ammunition to critics than it does to any apprehension over the likely quality and content of the answers. Questions such as those raised above are addressed in this study.

It would be difficult to exaggerate the importance of the shift in opinion leaders' attitudes in the late 1970s.[9] Today it is close to revealed truth to assert that the Soviet Union is bent upon posing the maximum possible prelaunch threat to U.S. silo-housed ICBMs—such is the inexorable logic of the Soviet fourth (and fourth and a half, perhaps)-generation ICBM program. Even though it was the United States that led the way with MIRV, the fact remains that for the better part of ten years it has been the U.S.S.R. that has been in the director's seat of the arms competition.

Public (and, hence, legislative) attitudes toward the level of defense expenditure, and large programs funded within that level, are not the product of careful defense analysis. The defense budget is both an expression of anxiety—how fearful are we?—and an affirmation of self-perception (U.S. defense posture is one important way in which Americans define themselves in regard to the outside world). Today the American public anticipates a dangerous decade ahead and shows no sign of being willing to be persuaded that "the danger is ourselves." Public opinion polls in 1980, for example, showed that 65 percent and more of the respondents would be more likely to support a presidential candidate who argued for an increase in defense spending than one who did not. The outcomes of the presidential and senatorial races in 1980 appeared to confirm this proposition.

Notwithstanding public anger and frustration with double-digit inflation, the years immediately ahead appear to offer the first genuinely permissive political environment for Soviet-oriented defense budget preparation since 1961-64. The middle and the late 1960s saw strategic and NATO Europe-oriented defense projects squeezed by the budgetary needs of the expanding Vietnam war— while defense spending as a whole in that period was under pressure from the Johnson administration, which was extremely reluctant either to abandon its domestic Great Society goals or to increase taxes to help pay for the war. The 1970s—until late 1979—experienced the post-Vietnam syndrome, wherein all U.S. defense projects suffered negative fallout from the nearly instant analyses of the alleged "lessons" of the Vietnam War.

President Nixon and Secretary of State Kissinger appeared to believe that the American public would not support a robust arms-competitive response to the Soviet strategic (inter alia) modernization drive—a belief that lend itself to rationalization through mistaken arms control aspirations.[10] Every administration in the 1970s (Nixon, Ford, and Carter) encouraged the view that there was an arms control alternative to an "unbridled nuclear arms race." Repeatedly the view was aired that there would be an arms control, or SALT, "dividend." No less eminent a strategist than the late Bernard Brodie encouraged the belief that an important, though not overriding, purpose of arms control processes was to save money.[11] It was, and still is, argued that SALT agreements lend "predictability" to the threat.[12] In late 1974, in the immediate aftermath of the Vladivostok summit meeting, Henry Kissinger claimed that the Vladivostok accords would place a "cap" on the arms race.[13]

Substantial U.S. self-restraint in the arms competition, over a 15-year period, has enabled the Soviet Union to close both the quantitative and the important qualitative gaps that previously had ensured its strategic inferiority. Intellectually, to the limited extent to which official attention has been devoted to the subject, probably no administration genuinely believed that defense planning problems could be alleviated greatly, let alone resolved, as a result of SALT regimes. Nonetheless, the very existence of the SALT process encouraged U.S. politicians and some senior officials to act as though strategic problems could be solved through SALT—and certainly to talk as though the SALT process would make a noteworthy contribution to the maintenance of a tolerable state of "essential equivalence."[14] The need to "sell" a SALT agreement leads, inexorably, to an inflation of rhetoric.[15] A president could hardly recommend a SALT treaty to the U.S. Senate if his language of justification included admissions that the treaty is not very important, technically, and that a treaty has been negotiated only because the U.S. political system requires evidence of "progress" in the arms talks.

The general public, though fearful for the 1980s—and apparently ready to take out a large defense insurance policy to help neutralize those fears—is still waiting to be told, by the national leadership, just how serious the danger may be. The Carter administration evidently was shaken in its innermost beliefs about the character of and rules governing East-West relations by the Soviet invasion of Afghanistan. But, in terms of policy substance, as opposed to gesture, the defense budgetary (outlays) increase proposed for fiscal 1981 was really only trivial. The promised 5 percent "real" increase would largely have been swallowed up by the increased cost of fuel. Harold Brown, as quoted above, uttered dire warning of possible impending military inferiority—yet the military posture he proposed to develop reflected, in the reality of force levels to be acquired ("the sharp end" of the defense establishment), the relaxed assumptions of 1977—early 1979. This is not to assert that Mr. Carter and Dr. Brown necessarily were wrong, but only to observe that the belligerent rhetoric of 1980 was not matched by either a clear statement of just what the president learned about the U.S.S.R. that he did not know prior to Christmas Eve 1979, or plans for the urgent development of a much more capable military posture.

While the U.S. government speaks the language of confrontation, real military muscle continues to be developed at a pace strongly suggestive of the operation of an informal "ten-year rule." (In August 1919 the British government—as an explicit premise for its defense policy—adopted the assumption that Great Britain would not be engaged in a major war for ten years: this "ten-year rule" finally was revoked in 1932, following the Japanese invasion of Manchuria.)[16] President Reagan <u>should</u> correct this asymmetry between declaration and programs.

The U.S. government, under Mr. Carter, appeared not to know

how to respond to the sharp recent decline in the tenor of Soviet-American relations, or to the (officially acknowledged) long-term relative decline in the Western end of the complex East-West military balance. Above all else, perhaps, the United States seemed unable to decide what, if any, were the political implications of the continuing adverse trend in the East-West military relationship. There was a dispute concerning the political implications of the following widely accepted "facts" about the Soviet Union:

Through most of the 1980s, the U.S.S.R. will have an unmatched, nearly total, prompt hard-target kill capability.[17]

The U.S.S.R. has achieved a clear superiority in theater nuclear weapons (embracing tactical, theater-operational, and theater-strategic).[18]

The U.S.S.R. retains its long-standing conventional superiority in, and bearing upon, Europe.

The U.S.S.R. has built an extra-European power-projection transport capability (airlift and sealift) such that Albert Wohlstetter's strictures in the 1960s concerning "the illusions of distance"[19]— in which he argued that in terms of transportation economics, or ton-mile costs, the United States enjoyed easier access to most of the Eurasian rimlands than did the centrally placed (with interior lines of communications) Soviet Union—are by and large no longer valid.[20]

The U.S.S.R. is in the process of building what is beginning to look ominously like a sea-control, as opposed to a sea-denial, navy. By the late 1980s the Soviet Navy will have heavily armed, nuclear-powered cruisers operating with genuine aircraft carriers.[21] Much of the prospective pattern of conflict at sea could be revolutionized by this development (a development, by the way, that has not superseded the heavy reliance on long-range, land-based naval aviation and attack submarines).

The U.S.S.R. has developed a network of overseas bases and arms depots (such as Syria, Libya, Ethiopia, South Yemen, Angola, Vietnam), together with allied "proxy" (and expendable) fire brigade forces, for rapid and massive intervention in regions once considered very remote from legitimate Soviet interests.[22]

The U.S.S.R. has continued to work hard to provide as best it can for the physical defense of, as well as the deterrence of attack against, the essential assets of the Soviet state. In a long-term and orderly program it has provided for the survival of its political and military leadership and major administrative cadres, is expanding shelter protection for industrial workers, and has created large and protected stockpiles of food, raw materials, and equipment. The leading Western authority on Soviet military thinking, John Erickson, has commented as follows on the Soviet approach to nuclear war:

> What was regarded as "unthinkable" in the West had to be rendered "thinkable" in the Soviet Union, for the survival of the system and its society was at stake. It was natural, therefore, that there should be a strong defensive input into this strategic evaluation and under- standable also (though incomprehensible in some Western circles) that any Soviet concept of "deterrence" could not be isolated from a defensive commitment: while Western commentators argued that "deterrance" and "defence" were incompatible, the Soviet argument insisted that "deterrence" without "defence" was as irresponsible as it was inane. [23]

The list above is not very controversial today. What is contro- versial is in the region of political implications.

A senior member of Mr. Carter's National Security Council staff, Victor Utgoff, was reliably reported to have acknowledged the adverse trends specified above, but to have argued that the United States simply could not compete effectively in the military field. He asserted that with double-digit inflation in the context of defense budgetary restraint, it was impossible for the United States to launch what would amount to a mobilization response to the Soviet military buildup. That line of thinking permeated the Carter administration's defense of the SALT II treaty. It was claimed that for all its imper- fections, the treaty both bounds (and renders more predictable) "the threat" through December 1985, and keeps the SALT process alive so that a SALT III could continue to restrain what the Soviet Union otherwise might choose to develop and deploy. [24] One underappreci- ated reason why the Joint Chiefs of Staff stood solid for SALT II was that the president impressed upon them that if SALT II failed to be ratified, spending on strategic nuclear forces would have to be increased greatly—thereby severely squeezing funds for other defense functions. It does not require a great logician to appreciate that this line of argument (which apparently was very effective) assumed a fairly constant level of defense expenditure. The administration ap- peared to be either unaware of, or indifferent to, the ancient maxim that the side that gives most evidence of wanting a treaty is almost certain to pay a negotiating price for that perceived fact.

The Soviet Union has yet to agree to any negotiated restraint that would likely have a net negative impact on its war-waging/war-winning ability. This argument does not carry the implication of Soviet villainy—far from it. The Soviets appear to believe that an arms control process is a useful adjunct to strategic planning—and prudent strategic planning for optimum military performance in defense of the Soviet homeland in time of war is nonnegotiable.

2

THE POLITICAL
AND STRATEGIC CONTEXTS:
UNDERSTANDING THE PROBLEM

To recognize the plain facts of an unfavorably evolving complex military balance is not synonymous with definition of the problem. Not many years ago, faithfully reflecting instant analysis of the apparent implications of the Vietnam experience, it was popular, or fashionable, to assert that military power had sharply declined in its utility. Small powers (though often they were close to being "regional great powers") such as Vietnam, Israel, Syria, and Egypt could, on this logic, "go to war" for advantage, or in fear, but the era of superpower (or nuclear-armed great power) military adventure was over: even if many soldiers, in the East and the West, did not recognize this alleged fact, their political leaders did.

As memories of Vietnam have faded, as Soviet proxy actions (most dramatically in Angola and Ethiopia) have occurred and succeeded, and most recently in the light (or half-light, perhaps) of the Soviet invasion of Afghanistan, the "inutility of military power, or force" thesis has declined dramatically in popularity. However, it retains a strong residual hold on Western imaginations and attitudes because it does reflect "our" cultural values, "our" deepest wishes, the character of "our" society, and the kind of international order in which "our" interests, by and large, can thrive.[1] Also, and this is not the least important of the relevant considerations, the implications of a rejection of the "inutility of force" thesis are awesome.

Unlike the Soviet Union, which is semimobilized for war today,[2] and is—and has always been (in good part for understandable geopolitical reasons)—a heavily militarized society,[3] the United States identifies grave social costs as a sure consequence of a noticeable measure of defense mobilization. Western insular societies, such as the American and the British, have an endearing, if ahistorical and foolish, tendency to believe, or desire to be

convinced, that life provides "happy endings." We wage "wars to
end wars"; we proclaim "eras of peace"; we are most reluctant to
believe that the SALT process is little more than a tactical diplomatic instrument for our manipulation; and we are exceedingly
unwilling to believe that "the detente process," launched with such
bravado in 1972, never had any substance worth mentioning. As
many commentators, domestic and foreign, have observed, the
United States appears to be condemned by its nature to swing somewhat violently and even erratically between policies of optimism and
accommodation and policies of rigid confrontation. Few phenomena
are more dangerous than a disillusioned liberal-optimist who has
just realized that there is evil afoot. Excessive optimism over Soviet
political intentions can easily be transformed into excessive
pessimism and alarm.

The U.S. political system is at something of a crossroads in
its foreign and defense policy deliberation. The deficiencies in past
policy have been at least half-admitted; Soviet military posture and
doctrine now are officially recognized fairly accurately on their own
(Soviet) terms, but the political meaning of the observed military
facts remains obscure. The Carter administration failed to reach an
internal official consensus on the definition of the foreign policy
problem (for example, are the Soviets [after Afghanistan] on the
march for "gold [oil] and glory," or does the current unpleasantness
in Afghanistan reflect nothing more sinister than Soviet politicians
sending in the military to "clean up the mess" on their Central Asian
frontier?), or believed that a very deliberate pace of military improvement, and alliance reinvigoration, would—over the long haul—
suffice to contain whatever Soviet leaders might wish to be about.
One thing, at least, was certain: the Carter administration did
"jump ship" from the long-popular notion that Western restraint in
the arms competition would encourage reciprocal restraint on the
Soviet side. In discussing the Soviet military modernization program
in Europe, Harold Brown ventured the following opinion:

> As far as we can judge, these developments [improve-
> ments in the group of Soviet forces in Germany in
> particular] have been substantially insensitive to changes
> in the magnitude of U.S. and allied programs for more
> than a decade. As our defense budgets have risen, the
> Soviets have increased their defense budget. As our
> defense budgets have gone down, their defense budgets
> have increased again.[4]

Some official "holdouts" in the Arms Control and Disarmament
Agency (ACDA) continued to resist Secretary Brown's characterization

of the _eigendynamik_ driving Soviet programs, but his factual obser-
vation reflected a most important official U.S. "discovery" about
the nature of the U.S.S.R. as an arms competitor. Some strategic
commentators made this "discovery" the better part of a decade
before the U.S. government, but their views were not taken seriously
at the time. A government believes what it wants to believe.

In the context of this discussion, it is important to note that
despite its evident uncertainty over the character of the Soviet
problem for the 1980s, the Carter administration did give every
appearance of having rejected the now-popular Soviet "window of
opportunity" theory. As with respect to the characterization of the
arms competition offered above (concerning a Soviet Union largely
driven by internally determined armament norms), the "window of
opportunity" theory has important implications for the MX ICBM
program. If it were believed that the probability of war will rise
markedly for period of five to seven years, beginning in 1981,
reflecting a much greater Soviet willingness to take risks—and
perhaps, logically, reflecting a sharp downward Soviet definition of
those risks—then the overwhelming defense budgetary priority would
have to be the provision of very near-term military capability.
While one might, and almost certainly should, continue and accelerate
the MX program, one would be casting around for a much "quicker
fix" than survivably based MX could provide. [5]

Stripped to its essentials, the "window of opportunity" theory
claims that the Soviet Union will enjoy a uniquely favorable period of
multilevel military superiority in the early and mid-1980s, and that
the West should anticipate Soviet political behavior to reflect this
fact. In extremis, this theory embraces a "now or never" thesis.
The most influential purveyor of this theory, though without the "now
or never" addition, has been Henry Kissinger. The following examples of his reasoning provide the flavor of the argument:

> The dominant fact of the current military balance is that
> the NATO countries are falling behind in every significant
> military category with the possible exception of naval
> forces where the gap in our favor is closing. Never in
> history has it happened that a nation achieved superiority
> in all significant weapons' categories without seeking to
> translate it at some point into some foreign policy
> benefit. [6]

With particular, though not exclusive, reference to the weakness in
U.S. hard-target counterforce prowess, Dr. Kissinger stated:

The consequence, to put it bluntly, is that in the 1980's
regional conflicts—whether deliberately promoted or
not—threaten increasingly to grow out of control unless
we drastically reverse the trend. We can not possibly
continue to gamble with inferior forces for regional
defense, a shifting balance in theater nuclear forces,
vulnerable land-based strategic forces, and invulnerable Soviet ICBMs without courting the gravest dangers.
The decline in relative power must be dramatically
reversed. [7]

There are several major weaknesses in the "window of opportunity" theory. First, it is dangerously military-mechanical in its
implications for foreign policy. Without challenging the claimed
facts concerning the military decline of the West, we do not know
how the Soviet leadership today, or in the mid-1980s, assesses or
will assess the risks and profit of confrontation or even of war
itself. [8] It is not self-evident that the Soviet advantage will be seen
by Soviet leaders to be a very directly exploitable advantage. In
other words, it is one thing to be worried by the impending fact of a
unique Soviet relative military advantage of five or seven (or longer!)
years' duration; it is quite another to assert as a fact that there will
be a Soviet "window of opportunity." The Soviets have ever shown
themselves to be prudent opportunists—they have never been
gamblers (adventurism is a crime against history in a political system with an ideology that promises inevitable eventual victory).
    Second, although there can always be a first time, politicians—
even in the Soviet Union—never "go to war," or take very large risks
of war occurring, simply because their general staff assures them
that victory is probable (how probable?). It is plausible that the
Soviet Union may attempt to exploit local circumstances in the early
to mid-1980s (in Iran, perhaps, or even Yugoslavia) in ways that
would have been ruled out as too dangerous in previous periods, but
those would be cases of intervention in the trend line Angola,
Ethiopia, Afghanistan.
    The theory holds that the Soviet Union has only a few years
(between five and ten) to reap the fruits of its military buildup dating
from the 1960s. By the late 1980s the currently planned Western
military modernization response will begin to shrink the Soviet advantage markedly. If one wishes to paint an apocalyptic picture, one
can envisage "preparing the United States for what they [SALT II
critics] see as judgment day with the Soviet Union." [9] It may seem
to be a rather weak and general argument, but it happens to be the
case—amply supported by history (which is evidence, of a sort, as
opposed to strategic crystal-ball gazing, which is not evidence of

any sort)—that responsibly governed countries—countries governed according to an ethic of consequences as opposed to an ethic of absolute ends—do not precipitate wars, or highly war-prone situations, simply because the military balance is relatively favorable. (But there is some dispute as to whether imperial Germany desired war in 1914 before the reform of the Russian Army could be effected fully, and Russian railroad construction shortened mobilization time dramatically[10]—but this author remains unconvinced on that issue.[11])

Third, it could be exceedingly dangerous for the West to believe that it has to survive, as best it is able, a five-to-seven-year period of maximum danger, following which all will be well. Despite the benefits of MX deployment (with an IOC of July 1986), and the availability of cruise missiles,[12] the general military picture for the late 1980s and beyond does not look bright. Too many Western commentators have chosen to focus upon "our" scheduled modernization programs, while neglecting those that are being pursued in the East.

The fact of the matter is that international politics, and the Soviet Union as a player therein, do not alter markedly over time— or lend themselves easily to analysis by periods. The Soviet Union is really imperial Russia with ideological trappings.[13] There have never been periods of "cold war" or "detente"; those were fictions invented by Western commentators with overly tidy minds and a poor geopolitical education. The Soviet Union is irrevocably committed to a long-term struggle with an outside world that is defined as hostile. The very long-term character of the struggle—once the excessive hopes of the immediate post-Revolution years had passed—permits, indeed requires, almost any amount of tactical flexibility. Accommodation with objectively hostile players (Nazi Germany in 1939-41; the United States in the early 1970s) has always been intended to forward Soviet aims in the long-term struggle. Unless the Soviet Union changed its nature (and its traditional Russian, not just Soviet, nature), Soviet-American relations could not evolve along a more constructive channel. American (and NATO-European) hopes in the early 1970s that the Soviet Union, in an "era of detente" and negotiations, gradually could be brought (bribed?) to play a stabilizing role in an emerging international order characterized by many cross-cutting linkages were simply naive.

So long as the United States is vulnerable to insubstantial foreign policy showmanship à la "detente-era" Kissinger, and now à la "Soviet window-of-opportunity" Kissinger, it will be next to impossible to engage in long-term defense planning worthy of the name. The MX program (and any similar project) can be imperiled by shrill alarmists who fear war in 1982 or 1983, and—a few months from now—by liberal arms controllers basking in the warm glow of a Soviet "peace offensive," arguing that the arms competition implications of MX deployment would be intolerable.

As it was, the liberal counteroffensive against the newborn "cold warriorism" of the Carter administration began with almost indecent haste. Leslie Gelb and Richard Ullman repeat a familiar critical thought:

> . . . it is by no means clear that the proposed MX
> intercontinental ballistic missile in its currently
> planned "racetrack" mobile mode would make sense
> if the SALT II treaty is scrapped. Freed from the
> limits of the agreed SALT ceiling, Mowcow could
> mount an attack with so many additional warheads as
> to make the mobile MX as vulnerable as the existing
> U.S. fixed-silo Munuteman.[14]

Thus, MX proponents are to be enticed into, or back into, the SALT II fold—within the general framework of business with Moscow more or less as usual. Close Washington watchers in Moscow probably broke open a bottle of champagne when they saw the Gelb and Ullman article. Gelb and Ullman gave no evidence of under-standing that it has been a declining level of conducting defense business as usual, for more than a decade, that has led to the current problem of the (prelaunch) vulnerability of the ICBM force. They do not recognize that it is fundamentally unsound to rest the viability of a major weapon program, such as MX, on a putative SALT agreement five or more years in the future.

This author is undecided as to whether there is more danger in irresponsible and shallowly reasoned Soviet window-of-opportunity theories or in no less irresponsible and blinkered urging that the West return, as soon as decently possible, to the transacting of East-West business more or less as usual. Both approaches imperil defense programs that have long lead times.

In the narrowest of technical senses, there is today very wide-spread agreement on the fact that the U.S. silo-housed ICBM force is becoming theoretically vulnerable to a Soviet counterforce first strike.[15] Even strong opponents of the MX system agree that the Minuteman force is in dire (again, theoretical) trouble. In the con-text of advocating ballistic missile defenses, for example, Stephen Rosen has asserted that "[t]oday, no one disputes that the threat [to U.S. silo-housed ICBMs] is real and imminent."[16] By contrast, one of the leading U.S. academic commentators on strategic theoretical questions was moved to entitle a paper "Why Minuteman Vulnerability Doesn't Matter"—which concedes the limited point being made here.[17] Stephen Rosen was too sweeping in his claim, in that there certainly are people who continue to dispute the reality of "the threat," although those people are in a rapidly shrinking minority. For reasons outlined briefly in Chapter 3, it is sensible to retain some skepticism on the subject of the current and future vulnerability of the silo-housed ICBM force. However, the ineradicable uncertainties of strategic analysis cannot absolve a defense policymaker from taking a prudently pessimistic (not "worst case") view of the issue. It is that person's duty, as this author has argued elsewhere, to pose very serious problems for any members of the Main Operations Directorate of the Soviet General Staff who are designing a briefing intended to persuade skeptical Politburo members that a strategic nuclear war could be "won."[18]

Despite the longevity of the matter, it remains the case that the U.S. body politic does not have a deep understanding of the strategic functions of its so-called strategic nuclear forces, while its appreciation of the net value of a land-based missile force tends to be less impressive still. Beyond that pejorative characterization, there are many analysts, officials, and politicians who have thought long and hard about strategic nuclear questions and have emerged from that process unconvinced that the United States either needs an ICBM force at all, or needs an ICBM force with the projected qualities of the MX. In other words, opposition to the MX program has several different bases. By way of a very broad-brush summary, there are several types of opponents:

Those who do not see the need for an ICBM force of any kind.

Those who favor an ICBM force, but not one with the putative hard-target kill capability of MX.

Those who favor an MX with hard-target kill capability, but question the timeliness of the extant program and doubt the viability of the proposed "survivable" basing mode—given the response options theoretically available to Soviet defense planners.

There are many variations on the above—most of the more important of which are treated in detail below. At this juncture in the argument it is important to recognize that although there is some closely reasoned opposition to the MX program—from its generic friends, as well as its generic foes—a great deal of the opposition, not excluding some of that which has been developed in a careful way, does reflect a measure of ignorance or insufficient information. Indeed, to be fair, there are some major aspects of the strategic rationale for MX that have, to date, escaped close scrutiny by almost everybody (these include the issues of the arms competition implications of the system, possible future SALT connections, and questions of targeting philosophy and related matters).

It is appropriate to observe that few strategic doctrinal issues have attracted such ill-considered commentary as has the value of strategic nuclear superiority.[19] As an overall characterization, it is accurate to note that the United States lost, or (more accurately) surrendered, strategic superiority at the end of the 1960s, embraced the concept of parity—variously and vaguely expressed as "sufficiency," "realistic deterrence," "essential equivalence"—as the guiding light for the 1970s, and now, at the beginning of the 1980s, has already entered a period that could, optimistically, be described as one of "tolerable (?) inferiority." The relationship between the perceived state of the strategic balance and the thrust and details of foreign policy is, necessarily, somewhat obscure.

However, the role of strategic forces in support of U.S. foreign policy is so obvious, and is of such long standing, that one can only wonder at the policies pursued by successive U.S. administrations. For various reasons, including geopolitics, comparative economic advantage, high-technology leadership, and domestic defense politics, strategic nuclear forces have been crucial as the backstop to U.S. foreign policy initiatives and responses worldwide.[20] U.S. strategic forces should not, and do not, have the sole mission of deterring a nuclear attack on the U.S. homeland. In a speech in 1979, Henry Kissinger—erstwhile advocate of, and apologist for, strategic parity—offered the following observations:

> [T]he change in the strategic situation that is produced by our limited vulnerability [the vulnerability of silo-housed Minuteman] is more fundamental for the United States than even total vulnerability would be for the Soviet Union because our strategic doctrine has relied extraordinarily, perhaps exclusively, on our superior strategic power. The Soviet Union has never relied on its superior strategic power. It has always depended more on its local and regional superiority. Therefore, even an equivalence in destructive power, even "assured destruction" for both sides, is a revolution in the strategic balance as we have known it.[21]

As Dr. Kissinger argued convincingly, the U.S. problem today is not merely that one leg of the strategic triad is losing its value for second-strike missions, but also that the United States very soon—pending the readiness of survivable MX—will face an inferiority that bears upon the most potent and flexible segment (the ICBM force) of that portion of U.S. military posture (the strategic forces) upon which, as Dr. Kissinger noted, the United States has placed uniquely heavy deterrent burdens. The ICBM is not "just another weapon" any more than the strategic forces constitute just one part of the combined arms team. Since 1953–54 the United States and its friends and allies have endorsed defense postures that, implicitly, embraced the idea of a planned deficiency in general-purpose forces. In short, we never intended to wage a very large war with the U.S.S.R. without having recourse to strategic nuclear weapons. Deficiencies in ground, tactical air, and naval forces relevant to war in Europe (as the focus, or "dominant scenario," of U.S. military planning) helped lend credibility to the threat of strategic nuclear intervention.

Through most of the 1950s and 1960s, U.S. strategic nuclear forces could have denied victory to the U.S.S.R. and, most probably, have limited damage to the U.S. homeland to such a degree that the concept of victory in World War III would have had some meaning. Strategic superiority, therefore, embraced the ideas of defeating the enemy and of avoiding the suffering of intolerable damage at home. With its acceptance, even glad acceptance, of strategic parity—supposedly registered in SALT I—the United States recognized the evolving facts of the strategic arms competition (which were not facts of nature, but facts of U.S. political choice). Unfortunately, perhaps, the rather obvious fact that strategic parity undercut the basis of U.S. overseas extended-deterrent duties somehow escaped official notice.

The long-standing anticipated inability of the United States and its allies to contain a Soviet attack in Europe has meant that unless the Soviet Union preempted strategically, it would be the United States that would need a credible (limited) first-strike capability. It would be the United States that would be compelled, by the logic of impending, if not actually consummated, defeat in the theater, to take the conflict to a higher level in quest of an improved outcome. In short, the role of U.S. strategic nuclear forces in a conflict around the Eurasian littoral would not, in all likelihood, be confined to that of a "counterdeterrent." Indeed, such a counterdeterrent should not be needed (save, possibly, with respect to the Soviet MRBM/IRBM force), since the Soviets would have no obvious incentive to raise the level of military action of the war so as to embrace the superpower homelands directly.

The case for retaining an ICBM force that is capable of hard-target kill and is survivable is a familiar one—though it is none the less valid for that. In the absence of such a force, even the semblance of essential equivalence would be lost. Without MX the United States would be lost. Without MX the United States would be perceived to be, and would be, strategically inferior. Such a United States would lack a "firebreak," and against this background a U.S. president would be unable to employ limited strike options.[22] Bereft of MX, the United States could not execute a militarily, and hence politically, intelligent war plan.[23] It is just possible that a United States with MX, though without homeland defenses on any scale, might be able to restore deterrence and terminate a conflict on acceptable terms; without MX a president's choices would be much constrained and the credibility of a U.S. strategic initiative, or even response, would be greatly weakened. By way of summary, the problem that is in urgent need of better understanding is that the decision to proceed, or not to proceed, with MX involves the quality of the deterrent effect on Soviet minds and decision processes that the United States should be able to enforce.

MX and the state of the U.S. strategic force posture as a whole, are not matters of arcane interest that are relevant only to almost unthinkable events. Instead, they are a very important element in the "peacetime," and crisis time, Soviet calculation of "the correlation of forces." It is well known that the Soviet Union has an operational, or battlefield, orientation in its assessment of force balances and political risks—it tends to military campaign analysis, as opposed to the U.S. policymaker-level tendency to see nuclear conflict in terms of bargaining theory.[24] In addition, testimony to the Soviet fondness for ICBMs is provided by the unambiguous evidence of their own actions: 70 percent of the total payload of the Soviet strategic forces resides in the deployed ICBMs (and probably rather more than 70 percent if silo reloads are included in the count).

Although there must always be some uncertainty as to "how much is enough," the United States can least afford to take the risk of military insufficiency with respect to its strategic force posture. Behind this judgment is the thought that overprovision of strategic nuclear muscle at least should ensure that the United States would always have the not incredible option of attempting to escalate its way out of a local defeat. Logically, if the strategic nuclear force posture is deficient, or is only very questionably adequate, the U.S. military posture as a whole has to lack integrity. Some critics of MX, and particularly those who adhere to some variant of a minimum deterrent doctrine, tend to turn this logic on its head. They argue that strategic nuclear employment would mean the end of history: Armageddon. Therefore, strategic nuclear forces are not "usable," although the distant (and really irrational) possibility that they might be used nonetheless will ensure that no country takes risks that might set in train a sequence of events that would evade careful political control. If "nuclear war is nuclear war," our—and their—fear of its occurrence should not be sensitive to the details of posture, doctrine, and plans. In short, the unprecedented destruction that nuclear war would entail virtually guarantees a massive stability at the strategic level of confrontation. In this view, probably the principal threat of such a nuclear war lurks in the "refined calculations of the nuclear gamesmen," to use McGeorge Bundy's words.[25] On this logic, it is just possible that usually sensible politicians might be persuaded by "clever briefers" that a nuclear war could be won.[26]

The minimum deterrer, or nuclear-war-is-the-end-of-history theorist, positively opposes those features of a strategic posture that would be of military-operational interest in the unwanted event (low CEP, high payload). It is important to recognize that some of the most vociferous opposition to MX stems from a small group of people who positively favor a strategic posture that could not wage a central nuclear war in a militarily intelligent fashion. They argue thus:

> . . . the sheer destructiveness of nuclear war has invalidated any distinction between winning and losing. Thus, it has rendered meaningless the very idea of military strategy as the efficient employment of force to achieve a state's objectives.[27]

This astrategic perspective upon strategic nuclear issues tends to be coupled with the view that the United States should attend, first and foremost (beyond, that is, to the improving of political relations with the U.S.S.R.), to improving its "usable" general-purpose forces. Although the United States could adequately deter for a quarter-century with general-purpose force inferiority and strategic nuclear superiority, the reverse (even if qualified to parity or essential equivalence) does not necessarily hold true. A Soviet Union that faced defeat in Central-Eastern Europe, or in the Persian Gulf region, could—and almost certainly would—escalate the conflict in search of an acceptable outcome. Unless the Soviet Union is denied the ability to devise plausible-looking plans for victory in a central nuclear war, no amount of correction of deficiencies in Western general-purpose forces can make for an adequate military posture.

Given the fact of his historical role as contributing architect to America's current strategic problems, it is fitting that a somewhat chastened Henry Kissinger should have the final words in this chapter:

> After an exhausting negotiation in July 1974 I gave an answer to a question at a press conference which I have come to regret: "What in the name of God is strategic superiority?", I asked. "What is the significance of it . . . at these levels of numbers? What do you do with it?" My statement reflected fatigue and exasperation, not analysis. If both sides maintain the balance, then indeed the race becomes futile and SALT has its place in strengthening stability. But if we opt out of the race unilaterally, we will probably be faced eventually with

a younger group of Soviet leaders who will figure out what can be done with strategic superiority.[28]

3

DEBATING MX

Underlying the simmering debate over the MX ICBM system lies an unresolved dispute over strategic doctrine. Many of the opponents of MX would continue to oppose the system even if there were no environmental impact, social impact, or budgetary or technical issues bearing upon the basing mode and the choice of basing area. These issues are important, yet secondary. MX proponents recognize the validity of questions concerning these aspects of the system and, consistent with strategic imperatives (operating and locating MX in such a way that the strategic purposes of the system are not compromised), are very willing to take constructive advice from any quarter so that the potential problems of social impact, land and water use, effect on local communities, and so forth, are solved as effectively as possible.

The people of Utah-Nevada should understand that the choice of their states for primary MX deployment was anything but an arbitrary decision. There are going to be sufficient real problems in need of cooperative resolution between the Air Force and those states, without false problems or unfounded suspicions being allowed to grow through a lack of communication. The problem here is not one of "selling MX"—that is exactly the kind of approach that invites (and almost deserves) local hostility. A proper approach to the people of Utah and Nevada should take the following form:

Frank admission that although MX deployment will bring some benefits, on balance Utahans and Nevadans may well judge the economic-social impact to be negative.

A reminder that the citizens of any potential MX-deployment state probably would feel the same way (except that in every other potential basing area, there are many more people, and consequently much more economic activity that might be affected).

A statement to the effect that the Air Force has a problem: for vital national security reasons (which should be spelled out clearly, in straightforward language—not in terms of "trust us, we know what is best") MX has to be deployed somewhere in the continental United States. What is the Air Force to do if it asks the citizens in every state, "Would you prefer that MX be deployed or not be deployed in your state?"—and in every state the answer was negative? Everyone can feel responsible and patriotic and say "I favor MX," then proceed to say, "but not in my neighborhood."

A very explicit outlining of just why the Great Basin of Utah-Nevada has been selected, out of six candidates, as the preferred basing area. It should be emphasized that the Air Force did not throw dice to select those states. Above all else, perhaps, it should be explained just what the strategic penalties or costs of deploying MX in some of the other basing areas would be. This honest approach should settle definitively the question "Why us?" That would not close off the local debate, or all local opposition, but it would certainly help to cut the ground out from under those people who are arguing that the Air Force could just as well, if not better, deploy MX elsewhere.

Probably most important, next to the answer to the "Why us" question, is for the Air Force to explain just how important MX is to national security. Sometimes there is an unfortunate official tendency: to treat the general public as though it consisted of simpletons who should be satisfied with very vague assertions (as opposed to arguments) concerning "the Soviet threat"; to resent offering detailed justifications—because of the belief that "we know best"; and to cut corners in explanations because the officials are so familiar with the concerns and arguments that they forget that strategic analysis, at almost any level of sophistication, is alien to the vast majority of people (which does not mean that they are incapable of comprehending it).

It is unlikely that the above approach would fail to be persuasive, though that possibility cannot totally be ruled out. Air Force briefings and discussions in Utah-Nevada, to date, have very much followed the lines suggested above. Some of the local (Utah and Nevada) opposition to MX deployment in their states is fueled, and even made strategically respectable, by genuine reservations over the strategic-technical wisdom of linear-track MX basing. There are some politicians who, almost regardless of their anxieties over, say, the possibly net-negative social impact of MX base construction in their states, would nonetheless be willing to risk their political lives by advocating local MX deployment, if they had confidence in the currently favored basing mode. The fact that they should have such confidence, yet some of them do not, clearly indicates the existence of an educational problem for the Air Force. In this connection detailed explanations are required concerning the desirability of horizontal shelters, the non-dependence for viability of the system upon SALT II (and SALT III) payload-fractionation sublimits, and the undesirability of digging vertical shelters in the existing ICBM fields with a view to rebasing Minuteman deceptively.

The above are contemporary highlights from the arguments of those who favor MX generically, are concerned about MX lead time (to the extent that they want what they believe will be a quick interim "fix" for the silo-vulnerability problem), and are looking for perfection. To the extent to which local and state politicians and the friends of MX who are determined to design a better system remain in the field contesting current deployment plans, generic opponents of the MX ICBM, on various grounds of alleged instability, need not risk very open combat with pro-MX forces. Although strong anti-MX policy-level arguments have been registered—placed on the public record, as it were—there has been scant need, to date, for the arms control "backwoodsmen" to do battle themselves. The MX program, almost as much for domestic political as for technical reasons, has been tripping over its own basing mode, and now—just possibly, if the problem is not handled well—over its basing-area difficulties as well.[1]

For different reasons most groups of actual and potential MX system debaters have betrayed some measure of nervousness at the prospect of entering a major policy debate on the subject. Quite unreasonably, though perhaps understandably, each group fears that it stands to be embarrassed. Also, the passage of time has posed considerable difficulties for each group. MX skeptics, in the mid-1970s, could express grave reservations over the reality, or at least the imminence, of the silo-housed Minuteman vulnerability problem. Soviet ICBM achievements since 1975 have destroyed the easy plausibility of a relaxed attitude toward the issue, meaning that MX skeptics are compelled, by and large very grudgingly (the issuing of mea culpa statements remains a rarity), to concede that the issue is indeed a problem—all the while stressing, to some degree quite properly, the theoretical character of the threat (given the large, and to some degree irreducible, technical uncertainties that would attend an attempt at a preclusive first strike against the entire U.S. ICBM force).

Those MX skeptics who now concede that a responsible defense planner does have a duty to do something to evade the acknowledged, though theoretical, danger have been mightily inventive in designing stability-innocent missiles, firing tactics, or defenses as substitutes for MX, and basing modes intended to evade some of the criticism that the various official schemes have attracted. Richard L. Garwin, for one very prominent example, has endorsed, and still appears to endorse, the following: cheap and simple active defense of silo-housed Minuteman; launch-on-warning (LOW) of silo-housed Minuteman;[2] and, for the latest brainchild, coastal-water-cruising, diesel-powered minisubmarines (with two, three, or four missiles strapped on the coaming).

MX proponents have the satisfaction of having predicted fairly accurately, by and large, the pace and character of evolution of the Soviet threat to American fixed hard targets.[3] However, some of the staunchest supporters of MX, and of a rapidly paced MX program, have come to qualify their support for the program as the rate of its evolution has, to be polite, to be described as stately. As late as 1976, the last year of the Ford administration, the IOC for MX was identified as 1983—a year that seemed, albeit somewhat relaxed, likely to be responsive to Soviet threat development as then it was understood. If anything, Soviet MIRV deployment and CEP improvement have proceeded at a pace brisker than anticipated even by typically conservative (in a political sense) threat estimators. So, by 1980—81 some MX proponents were wondering whether a program that looked very attractive, and that they had supported strongly in 1974—76, remains sufficiently responsive—with a July 1986 IOC and a (calendar year) 1989 FOC. These doubts have fueled a search both for "quicker" (than MX) fixes, and for "interim" fixes, pending the availability of MX. The advocacy by Paul Nitze and by William Van Cleave,[4] for example, of an ALPS-MAPS (alternative launch-point system; multiple aim-point system)—entailing the (immediate) digging of new "silos" in the current Minuteman fields—has been motivated by a sincere concern lest the MX program lead time will not produce enough survivable ICBM firepower soon enough.

It is worth observing, in passing, that the attraction even of "quick fix" ideas is subject to erosion by time. Provided the MX program schedule does not slip from 1986, as each year passes, "quick fix" solutions offer less benefit. Also—and this is a point that their advocates do not deny—in a defense political climate as volatile as the American, serious strategic analysts have grounds for a residual nervousness concerning the real costs of "quick fixes." In defense, as in everything else, there are no free lunches. The actual price, considered multidimensionally, of "quick fixes"—notwithstanding the impeccable motives and careful analyses of their authors—could prove to be exorbitant. Generically, a quick fix may prove to be neither as "quick" as assumed nor as effective a "fix" as hoped—

while implementation of the intended "quick fix" might fatally compromise the prospects for an enduring solution to the problem in question.
The whole subject is part technical and part political-policy judgment.[5] To repeat a theme aired earlier, if one really believes that
an acute East-West crisis, military confrontation, or even war is
very probable over the next couple of years, then one has no
responsible choice other than to attempt to "surge" U.S. combat
capability, almost regardless of imperfections in the "quick fixes"
effected and of possible negative consequences for other defense
programs with a much longer lead time. The United States will not
be awarded points, let alone a favorable decision, in a crisis or a
war on the basis of the predicted future performance of the wonderful
weaponry that is under development.

The present low level of the policy debate over the MX program
flows in good part from the fact that many pro-MX people have felt
inhibited from debating a system whose technical details were still,
to a degree, uncertain,[6] while many anti-MX commentators have
seen no need to raise policy questions very forcefully when the system
appeared to be in trouble on technical and local political grounds. A
further reason for the reluctance to raise policy-level issues concerning the implications of the system is that no debating group has
been confident that it stood to gain thereby. In the opinion of this
author, such an assumption, on the part of MX advocates at least,
has been a serious error. However, as already noted, the scarcity
of genuine strategic debate in the U.S. defense and arms control community seems to be an enduring consequence of the structure of
government, of an American "national style,"[7] and—most prosaically—
of the lack of qualifications and interest of many of the potential
debating contenders.

MX advocates have tended to be nervous of making a case for
the MX missile, as opposed to "survivable ICBMs," because a focus
on the missile has to lead to a discussion of why large payload (for
U.S. ICBMs) and small CEP (high accuracy) is deemed to be desirable.
Logically, this leads one into a discussion of deterrence focused upon
Soviet fears and vulnerabilities regarding MX as a war-fighting
instrument. It appears to have been judged, until very recently, to be
the case that opinion leaders in the U.S. political system were not
quite ready for forceful explanations of why "a good defense is a good
deterrent," and of why there is no contradiction between deterrence
and defense. In addition, no U.S. administration in the 1970s endorsed the evolution of a U.S. (hard-target) counterforce first-strike
capability—although James Schlesinger came close to so doing. Harold
Brown, for example, at least prior to his explanation of the "new"
targeting doctrine outlined in PD 59 of July 1980, tended to defend MX
(inter alia) as a second-strike counterforce equalizer.[8] Thus, it has

been feared that development of a strategically rigorous story in
defense of MX could accomplish the following:

It might take the Air Force, publicly, some way beyond official
strategic thinking.[9]

It might have a negative political effect because of its unavoidable
discussion of escalation options.[10]

It might risk widespread misunderstanding, in that, as stated above,
the "defense for deterrence" thesis is not as well appreciated, or
liked, in the United States as it could be (or as it is in the U.S.S.R.)[11]

  This author believes that these fears have some declining basis
in fact, but that the "fact" in question urgently needs to be changed
through strategic education. After all, the Department of Defense
cannot pretend that 200 (10 reentry vehicle [RV]) MX ICBMs, if
launched in a first strike against silo-housed Soviet ICBMs would
not achieve a very substantial disarming effect. Moreover, MX
critics, particularly once the deployment mode and basing area
issues finally are resolved, will not permit the Department of
Defense to pretend that there are no doctrinally fundamental matters
pertaining to the system.

  MX opponents appear to be nervous of raising strategic policy-
level questions about the system because, in doing so, they invite
debate on ground that offers no sure footing. Leaving aside the finer
points of stability theory, they would risk being embarrassed by the
plain facts of the rapid evolution of the Soviet hard-target kill capability. They are in the position of asserting that it would be "destabilizing" (a concept as vague, useful, and much loved as it escapes
rigorous definition)[12] for the United States to deploy a nearly total
threat to the prelaunch survivability of the Soviet ICBM force by
1989, when the Soviet Union would have deployed such a threat
against American ICBMs by 1980−81. Nonstrategic analysts, un-
versed in the theology of stability theory, should be expected to
wonder why the United States should feel constrained from doing in
the future what the Soviet Union is doing today.

  With respect to the debate between defense and arms-control
professionals, generically anti-MX commentators are caught without
a tried and trusty road map. By and large they have jumped ship
from substantially unqualified endorsement of the MAD thesis. They
have recognized, with few exceptions, that the thesis lacks authoritative Soviet endorsement;[13] that MAD was always a budgetary and
force-sizing control tool, rather than a strategic doctrine;[14] and that
MAD has nothing of operational interest to offer a strategic planner.
However, once one backs off a little way from the proposition that

nuclear war is an unqualified holocaust, where does one go? The anti-MX debater of strategic policy questions has problems.

He/she cannot, in good conscience, oppose the idea of survivable ICBMs (though he/she can argue that a strategic-forces "dyad" would be good enough).

He/she knows, if sensible, that tomorrow's weapon system will be more capable, unit for unit, than today's—and that technological progress (weapon performance) cannot be halted.

He/she is obliged by common sense to endorse the idea of flexibility and selectivity in strategic targeting. Even though convinced that World War III would be an uncontrollable holocaust, he/she cannot advocate a policy that would, by choice, ensure a holocaust. [15]

He/she is aware that a U.S. decision to place much, or most, of the Soviet strategic force posture off targeting limits would not be reciprocated by the Soviet Union. [16]

In essence, the opponent of MX who argues that MX will assuredly and needlessly promote arms race and crisis instabilities is advancing a case that, given its internal logic, had some relevance to the real world, as it was (falsely) perceived, of the late 1960s and the very early 1970s. That case cannot accommodate the strategic history of the 1970s or the problems foreseeable for the 1980s.

Finally, there are three elements in the MX debate that tend to detract from proper appreciation of the central strategic issues, and that constitute traps for unwary debaters: the problem of overstatement, the "threat of the week" phenomenon, and strategic uncertainty.

First, overstatement is an inevitable feature of any defense debate. Analysts become partisans and tend to assume the role of lawyers with a "brief" for their client's position. In the real world, alas, there are no "analytic policemen,"[17] and there are relatively few penalties for exaggeration, ignorance, or perjury. One is reminded of the analogy of the glass half full or half empty. MX proponents are correct in asserting that it is an impending strategic fact of great significance that the ICBM leg of the strategic-forces triad will soon, theoretically, be liable to prelaunch neutralization. However, MX skeptics also are correct in asserting that such an asymmetrical vulnerability, to the disadvantage of the United States, need not mean that the United States would be strategically inferior in any important respect—provided one adheres to some variant of an urban-industrial punishment focus for one's theory of nuclear deterrence.

Minuteman vulnerability should mean, in the view of this author, that the United States will face unprecedented risks of political and military defeat, but that view cannot be stated as though it were an axiomatic truth. MX advocacy, when overstated, lends itself to ambush by skeptics who point, quite rightly, to the judgmental elements in the overstated advocacy. Similarly, strongly worded attempts to belittle the significance of unfavorable asymmetrical silo vulnerability lend themselves to the charge of strategic imprudence or irresponsibility. There is a danger that truth and genuine unknowns may become early casualties: that silo vulnerability is a guarantee of U.S. (and U.S. allies') humiliation or defeat, or that silo vulnerability does not matter.

Second, ingenious defense analysts can always invent, sometimes literally on the back of an envelope, a threat to which the weapon system in question should be fatally vulnerable. This "tiger team" or "threat of the week" phenomenon, generally is a healthy feature of the defense community. No one's preferred weapon, tactic, or strategy should be accorded a free ride, safe from criticism. However, the merits of a wide-ranging intellect can be less obvious in the context of a hard-fought adversarial debate over a particular weapon system. The restless scientific imagination can deploy space-based laser BMD systems, can find weapon deployment signatures (usually unspecified), can provide real-time reconnaissance and maneuvering reentry vehicles, and so on. Superthreats to MX are invented quite easily—but they tend to require a quality of C^3I that is improbable, a degree of faith in military dexterity that real-life politicians are unlikely to hold, and a Soviet General Staff willing to ignore the long-standing precept that prominent among the virtues of a sound military plan is the quality of simplicity.

Militarily or politically, implausible "threats of the week" can do damage to a program out of all proportion to their inherent merits. Above all else, perhaps, such claimed threats can be impossible to disprove definitively. For example, to an analyst in 1981, destroying MX missiles in 1995 with Soviet space-based lasers is a difficult target to debate.[18] In answer to the direct question "Could the Soviets do that in 1995?" an honest MX proponent has to say either "Yes, but . . . ," or "Perhaps." For some audiences the case of the MX proponent might be lost once he/she concedes the distant possibility that the ingenious threat designer might be right. The audience may be more impressed by the fact of concession that it is by the qualification that it is a "distant possibility." Needless to say, perhaps, the ingenious threat designer tends to discount those ingenious threats that are leveled at his/her preferred weapon system.

Third, the debate over the MX ICBM system has begun to attract some quite extreme use of the fact of strategic uncertainty. In an article Robert Kaiser and Walter Pincus, with journalistic license, report a fictitious briefing-and-answer session in Moscow, in the course of which the Soviet leadership is made to appreciate that the theoretical vulnerability of U.S. silo-housed ICBMs does not translate into any very useful war-waging options.[19] It is certainly true to claim that defense commentators very often—indeed, almost habitually—assume that very complex weapon systems, C^3I, and man-machine organizations at many levels of government will do what, in theory, they are intended to do. It is scarcely less true, or less of a truism, to note that very, very few defense professionals (inside and outside of government) are unaware of the extent to which military capability assessed in peacetime could differ from wartime performance. One may not embrace "Murphy's Law," but "the foul-up factor" is universally appreciated (witness the "rescue" mission to Tehran).

From time to time it is useful for the defense community to be reminded, in a systematic way, of the strategic uncertainty that is inescapable in regard to the estimated quality of actual performance of its military output.[20] Uncertainty is of different kinds (including technical: how will U.S. missiles perform when fired over operational trajectories?; or political-operational: what will a U.S. president choose to do? how will the Soviet Union respond?), and is pervasive and probably cumulative. Defense planners make proper statistical allowance for probable technical failure, but—particularly in the realm of strategic nuclear warfare—they are attempting to comprehend a subject that is beyond their experience. The ultimate uncertainty, beyond the specific lists that can be assembled, reflects a fear that there may be something about the conduct or effects of nuclear war that we do not know about at all.[21]

If one attends diligently to the question of strategic uncertainty, it is not too difficult to persuade oneself that, virtually no matter how the strategic balance stands, every prudent politician should lack adequate confidence that any promise of military victory really could be delivered. This line of thinking could lead to the conclusion that silo vulnerability does not matter—it is a vulnerability only in theory (regarding untried [in an operational context] weapon systems). Very few policy commentators have been so bold as to dismiss alleged strategic problems by means of reference to the uncertainty factor. But many commentators, particularly in debating MX, have resorted to the uncertainty argument as a strategic "makeweight."[22]

Unfortunately, a defense planner cannot translate very many strategic uncertainty factors into useful precepts. He/she cannot assume that the Soviet ICBM force will fail catastrophically in its hard-target kill mission; nor has he/she any basis for assuming that the Soviet leadership would, or would not, launch its ICBM force on receipt of strategic or tactical warning. In short, once one proceeds beyond noting the possibility of catastrophic technical failure, and the possibility of political leaders behaving in a strategically irrational manner, one lacks any guidance as to how to render "strategic uncertainty" concerns useful for defense planning.

Strategic uncertainty is almost a joker or wild card, to which one can assign an arbitrary value for the convenience of the moment. Virtually every strategic weapon system and strategic C^3I system, in principle, is vulnerable to the charge that it would not "work" in a nuclear war[23]—moreover, to tone down the charge, it is not always obvious on what basis an official can even assert that a particular weapon system is judged to be very likely (how likely is that?) to be able to perform its mission. It is probably reasonable to insist that the burden of proof lies upon the weapon proponents, rather than the skeptics, but how can defense planners seek to persuade the skeptics about a class of military equipment (for nuclear war) that cannot be tested rigorously short of the actual event that the equipment is designed to help deter?

It is probably useful to remind readers that strategic uncertainty, considered very broadly, is a positive phenomenon for international security. Potential first strikers are necessarily uncertain of the merits of their plans and the performance of their weapons. Some MX program critics are skeptical of the ability of the system to withstand a Soviet saturation attack—but how should the problem look in Soviet perspective?

PART II
THE MAJOR QUESTIONS

The debate over a follow-on ICBM to Minuteman III, its chosen basing area, and its selected deployment mode has been sufficiently protracted and detailed that it is possible to summarize the major questions and provide at least an outline of appropriate answers—all with considerable confidence that little of real significance will be omitted. The complex nature of the debate and the partially overlapping—and certainly mutually dependent—character of many of the issues have driven this author to identify, and seek to answer, separable questions that, in toto, constitute the lifeblood of the MX debate.

The questions and answers that follow are designed in as objective a spirit as possible. "Straw men" are not attacked, and objective characterization of proposals and arguments with which this author happens not to agree is attempted.

4

MINUTEMAN VULNERABILITY AND
THE FUTURE OF ICBMs

IS THERE A MINUTEMAN VULNERABILITY PROBLEM?

Commentators as diverse as Harold Brown, Henry Kissinger, and McGeorge Bundy agree that there is.[1] There is an absence of agreement on whether that vulnerability is a more, or a less, urgent problem, just as there is no consensus, as yet, on whether MPS basing of an MX ICBM constitutes the most desirable solution. However, the defense and arms control community does agree that the silo vulnerability problem is a real one. In other words, the MX ICBM, survivably based, would constitute a candidate solution to a genuine problem. In principle, if a great many things went "right on the night," the U.S.S.R. should—today—be able to reduce the Minuteman-Titan ICBM force to an unacceptably low level by means of a preventive, or preemptive, first strike. As noted in Chapter 3, there are many reasons for Soviet strategic uncertainty over the quality of likely performance of their offensive systems—employed on so massive a scale, in so carefully integrated a fashion, for the first time in history. However, if the U.S. defense community ignores the possibility of catastrophic Soviet technical or operational failure, and instead assumes modest but solid Soviet competence, then the theoretical first-strike vulnerability of U.S. silo-housed ICBMs simply has to be conceded.

It is worth recalling that as recently as the mid-1970s, Soviet ICBM CEP was predicted to remain too high for many years for a really serious threat to be posed to Minuteman.[2] A high CEP meant that more than two warheads would be needed for high confidence of a kill against a silo-housed ICBM, while the allocation of more than two warheads to one aim point (always presuming a warhead inventory of adequate size) would raise the problem of offensive fratricide to successively higher and higher levels.[3] Even if the Soviets were to

believe that their planning and their equipment would be good enough
to permit them to meet the very exact time-of-arrival "window"
requirements of, say, a three-on-one or even a four-on-one attack,
they could hardly be highly confident of technical success. Overall,
deterrent effect is very much a product of levels of confidence—or
the lack thereof.

Although there is widespread agreement that there is a problem, consensus breaks down when one attempts to define the precise
character of the problem. Many of the people who grant that it is
undesirable to have an ICBM force that is vulnerable to preventive or
preemptive attack, are disturbed very largely for the somewhat general reason that they judge a strategic-forces triad to be more helpful
than a dyad in promoting what generically is termed "stability." They
do not believe that the U.S.S.R. would dare to attack the U.S. ICBM
force alone, given the enduring prelaunch and penetration survivability
of SLBMs and manned bombers/ALCMs. Therefore, they view the
silo-housed Minuteman vulnerability problem as a challenge to that
multiple redundancy in strategic capability that they deem to be
valuable for deterrence. In addition, there is nearly universal
recognition of the fact that each leg of the strategic forces triad
functions synergistically with the others for the enhanced security of
them all. If the ICBM force were to be phased out altogether, Soviet
on-line strategic assets would be able to concentrate upon the simpli-
fied threat profile presented by a U.S. strategic dyad, and Soviet
research and development would be able to narrow its focus accord-
ingly. In short, it would be an unhealthy trend.

To acknowledge the Minuteman vulnerability problem does not
imply anything in particular concerning the needed urgency in finding
a solution, or the character of the solution. If one believes that
general war inevitably would be a nonsurvivable holocaust—in short,
if one denies the plausibility of James Schlesinger's or Paul Nitze's
scenario of limited central war[4]—then one is unlikely to be very
perturbed over the strategic, or perhaps even political, meaning of
silo vulnerability per se. Manned bombers alert on runways in
CONUS and ICBMs in silos, whether vulnerable or not, function
synergistically for survival. In other words, one may grant the
vulnerability problem, endorse a determination to find a solution,
yet be fairly relaxed concerning the lead time for the on-line provision of the chosen solution. As McGeorge Bundy has argued
repeatedly over the years, politicians—unlike defense professionals—
are very likely to be deterred by the loss of a few cities.[5] While it
is desirable, according to the prudent version of this logic, that the
large-scale destruction of cities should be a capability of each leg
of the triad (hence the desirability of attending to Minuteman vulnerability), the continuing ability of two legs of the triad (and even a

disorganized small "rump" of the third leg) to wreak assured city destruction should ensure that the silo vulnerability problem does not lead to any overall loss in deterrent effect.

It is interesting to note that the redundancy-for-assured-destruction case for solving the vulnerable silo problem leads Bundy, for example, to a position where he should have been friendly to the Carter administration's proposed lead time on MX—in somewhat uneasy company with many defense professionals. Interestingly, Bundy's rationale for a follow-on ICBM (or some other third leg for the strategic force posture) is considerably (though not totally) different from that expressed officially by the Department of Defense. Bundy has expressed his concerns as follows:

> It is a renewal of insurance, not an increase of destructive power, that is needed. The currently proposed MX missile may well be bigger than what is needed by this standard. [6]

and

> Because the deterrents on both sides are so ample, there is a large margin of safety, and it is reasonable to set our forces for example above the danger level to ensure general public confidence. [7]

As may be deduced from this discussion, although public recognition of Minuteman vulnerability as a problem has been essential, and is now accomplished (albeit belatedly), that recognition constitutes but a nursery slope on the path of general public understanding of the strategic case for MX/MPS.

DOES THE UNITED STATES NEED AN ICBM FORCE?

Some people have responded to the imminence of silo vulnerability by defining the impending condition as a challenge rather than a problem. They argue, not without some justification, that the strategic triad was more the accidental product of interservice politics and the fact of largely parallel, successful strategic development programs in the late 1950s, than it was the intended (and intended to be enduring) outcome of rational defense planning. [8] The argument takes the following form: silo-housed ICBMs have served the United States well, if very redundantly (given the survivability of SLBM tubes and manned bombers), for 20 years. However, no weapon system can be survivable forever; the ICBM, like the horse cavalry, has had its day.

It is argued, further, that although ICBM prelaunch survivability might be extended through some combination of deceptive basing and active defense, fundamentally the ICBM would be engaged in a losing competition against the quantity and quality of the threat (which happens to comprise—for the 1980s, at least—the Soviet ICBM force). Generically, this argument holds that the United States, in pursuing MPS basing for MX (or any other ICBM), will be fighting the tide of technological history. The MX system, on this logic, will prove to be incredibly expensive and ultimately ineffective.

This author is not unfriendly to the thesis that years of incremental change in strategic capability (for example, in warhead numbers, in CEP, in yield-to-weight achievement, in reliability) eventually bring forth a situation where the viability of a whole class of weapons needs to be reexamined. In principle, certainly, one has at least to entertain the possibility that ICBM/MPS would constitute an incremental response to a threat that could not be so contained. However, as later discussion will explain, there is excellent reason to believe that the heralding of the demise of the ICBM is, at best, premature. This author believes that the ICBM is no more obsolescent than is the armored fighting vehicle—enthusiasts for antitank guided weapons are no more persuasive than are those who advertise the total hard-target counterforce competence of the U.S.S.R., regardless of the scope of U.S. ICBM survivability programs.

Three points need to be made on the "tide of technological history" front. First, there is a constant dialectic between the offense and the defense.[9] There is no law of military history that states either that the offensive challenge will always succeed or (ergo) that the defender is well advised to abandon the competition when the adversary fashions what could be a comprehensive threat. If the Soviet offensive threat to the U.S. ICBM force triumphs in the arms competition in the early 1980s, it will be because the United States chose not to compete. There is no good reason to believe that the United States cannot compete successfully with the Soviet offensive challenge.

Second, defenders of a U.S. follow-on ICBM program are not, and should not be, arguing that the United States either should, or will, be able to maintain an adequately survivable ICBM force forever. Instead, they are saying that for the defense planning period of current interest (say, to the year 2006),[10] MX ICBM capability is deemed to be essential, and that adequate survivability against plausible, and even some implausible, threats can be ensured through that period.

Third, this author is suspicious of an argument that holds that the "tide of technological history" somehow is running against U.S. ICBMs, but not against Soviet ICBMs (not, at least, until well into the 1990s—if and when the United States in hard-target kill-capable

SLBMs [say, the Trident II]. In the interest of what this author means by strategic stability, [11] the crisis of confidence in ICBM viability, if it has any reality, should be mutual.

Deferring, for the moment, detailed discussion of the question of the viability of ICBM forces, it is necessary to summarize the reasoning that generically pro-ICBM and anti-ICBM debaters are wont to display. Defenders of a U.S. ICBM capability deploy a wide range of different arguments—not all of which are universally popular (because they reflect particular, and controversial, beliefs concerning strategy).

The following arguments are presented for an ICBM force:

It is the most accurate means for striking promptly at time-urgent targets.

It has uniquely reliable C^3.

It is always "on station" (that is, has very high operational readiness).

To date, at least, it can be threatened only by a very large strike against American soil (which should be a uniquely deterring prospect).

It enjoys a preeminence of respect in Soviet eyes (witness Soviet practice).

It provides cover, through attack-timing complications, for a manned bomber/cruise missile carrier force.

It diversifies the threat faced by the Soviets.

The ICBM is the counterforce weapon par excellence. Although penetrating manned bombers and cruise missiles can deliver warheads with yield and accuracy combinations that provide a hard-target kill probability comparable with that of an ICBM RV, they are both relatively slow to complete their mission and would be opposed by air defenses. Pending the deployment of substantial BMD, there is a prompt certainty about an ICBM strike that cannot be claimed for other kinds of strategic capability. Whether or not one is interested in the ability to destroy hard targets (military, political, economic, and even cultural-symbolic) is a matter of doctrinal persuasion. Indeed, one might follow James Schlesinger and Henry Kissinger in holding the opinion that although a very substantial hard-target counterforce capability is an option of only limited interest to the United States, we should not be prepared to permit unilateral Soviet acquisition of such a capability. [12] There is a coercive potential to a unilateral counterforce capability with which the Soviet Union should not be trusted—notwithstanding the bizarre features of the pertinent action scenarios.

At root, the case for ICBMs of the proposed MX generation (the U.S. fifth generation) has to be assessed in terms of preferred strategic doctrine. A very capable ICBM lends itself to employment for traditional military purposes (to defeat the armed forces of the enemy)[13] and, of course, for the more novel purposes of intrawar bargaining (to defeat the political will of the enemy). In the opinion of this author, if the United States were unilaterally to abandon its ICBM force, it would thereby concede operational strategic superiority to the U.S.S.R. Those words are not employed as political rhetoric, nor do they refer to politicians' perceptions of the implications of static indicators of strategic capability.

To be specific, without an ICBM force the United States could not execute an intelligent SIOP, or even pre-SIOP limited nuclear options (LNOs). At a stroke the United States would have deprived itself of the ability to strike promptly and reliably at Soviet hard military and political control targets—precisely the target sets that the U.S. defense community has identified as being the most critical in terms of prospective deterrent effect.[14] Manned bombers and cruise missiles could destroy those targets, but it would be near-suicidal to assign them such missions in the absence of massive active-defense suppression strikes. It is true to claim that a United States bereft of its ICBM force could easily reduce much of the urban-industrial Soviet Union to rubble, but it also happens to be true that an ICBM-bereft United States should be convincingly self-deterred from ever exercising such an option.

To summarize, a survivable ICBM force (say, to the 50 percent level identified as desirable for MX) should function as an escalation "firebreak." Its very existence should be a nightmare for Soviet defense planners. They cannot ignore such a potent U.S. capability—yet they cannot attack it in expectation of net gain. All the while, that survivable U.S. ICBM force should yield some freedom of strategic action to a desperate U.S. president. Readers may be assured that although they, personally, may see little to recommend a strategic forces triad that includes an ICBM force, there is a well-developed and very widely endorsed case for such a force. The real momentum behind the MX program is not provided by bureaucratic-service self-interest, strategic habit, or the profit motive in defense industry. Admittedly, all of these factors are present. A portion of the U.S. Air Force is in the missile business and wishes to stay in that business. Also, it is true to claim that strategic habits die hard and that the U.S. defense community has become used to a strategic forces triad and to an ICBM force. Finally, it cannot be denied that there are profits to be gained from building the MX missile and its basing structure. However, the MX program is proceeding for none of those reasons: they are facts, but they are not driving factors.

There is an apparent "mad momentum" to much bureaucratic
activity (including some weapon programs and some arms control
activities), but the MX/MPS ICBM program is an almost uniquely
example of such. It happens to be a fact that by way of very sharp
contrast with Soviet standard operating procedures, the United
States operates a feast and famine cycle in the procurement of major
weapon systems.[15] The United States, rather than "tinker" with a
system from year to year, has tended to procure genuinely new
weapon systems (step-level jump "improvements" over the previous
generation) in a surge production mode, and then to debate the next
step-level jump over a multiyear period. None of the standard
military-industrial complex hypotheses fare very well in the context
of the history of the MX program to date. The Carter administration
aborted the B-1 in 1977; it could have done likewise to MX in 1978
or 1979. If any "mad momentum" is to be blamed for the green light
given MX development in 1979, it was the perceived domestic political
requirement deriving from a "mad momentum" of arms control
(SALT II).

5

THE MX/MPS PROGRAM

IS MX/MPS THE CORRECT SOLUTION?

The most honest and accurate answer that can be provided in response to this question is that MX/MPS is the best solution to the vulnerable silo problem that the defense community has been able to provide—given the many political constraints. Moreover, and this modifier is critically important, MX/MPS promises—on current evidence—to be good enough. If the U.S. government could do as it pleased, if it had no domestic opinion to consider, then it is possible that the follow-on system to silo-housed Minuteman III would be very different from the MX program as configured at present.

To be specific, one ideally would select a follow-on ICBM basing mode that should be "threat-size independent." A structural, theoretical weakness in MX/MPS is that it provides an adversary with a finite (if very large) target set. Ideally one would deploy modestly dimensioned ICBMs in trailer trucks and would conceal them within the traffic flow of the nation's highway system. However, for reasons of democratic politics—public reaction and security—a truly road-mobile ICBM deployment simply is not feasible.[1]

Since the late 1950s (since before Minuteman I came "on line"), the U.S. Department of Defense has considered well in excess of 30 different basing modes for ICBMs. Minuteman I originally was intended for railroad flatcar deployment, but—since the late 1950s—scarcely any interesting land, lake, canal, pond, airborne, or coastal-water basing option for ICBMs that ingenuity could devise has lacked for a study (often many studies) or a persuasive-sounding group of analyst-advocates.[2] If there is one thing above all others for which Americans are justly renowned, it is the design of "engineering fixes." The ICBM survivability problem, to summarize, has been studied, and restudied, exhaustively. Detailed issues pertaining to the MX missile and the "linear-track, plow-out, horizontal shelter" deployment are treated below. However, critics of MX/MPS should be aware both of the quantity of study that has been devoted to the survivable ICBM issue, and of the political constraints that inevitably have had an impact upon some of the engineering details of the program as it has evolved.

The truth of the matter is that probably half a dozen basing modes would be "good enough" to ensure ICBM survivability in the late 1980s and the 1990s. Every candidate basing mode has had distinctive merits and demerits, many of which, in practice, probably would prove to be trivial. It is the belief of this author that an MX/MPS system[3] is quite good enough to deter Soviet competitive challenge; to defeat Soviet competitive challenge, if offered; and to provide sufficient survivable ICBM payload to complete the offensive missions that U.S. targeting schemata may lay upon it.

Whether one likes it or not, the U.S. defense community is constrained to design a future ICBM basing mode that is SALT-compatible and does not generate politically intolerable public interface. MX/MPS is verifiable by (Soviet) national technical means (with some intelligence assistance from Aviation Week and Space Technology),[4] preserves all the essential attributes of a silo-housed ICBM force, and is minimally disruptive of domestic society (given that the prospect of the active defense of existing ICBM silos is held to raise intolerable problems).

With respect to the modifying consideration offered immediately above, this author is tempted to suggest that BMD of the existing ICBM facilities is an attractive route that the United States could traverse. However, it is a technical fact that the active defense of a nondeceptively based ICBM force requires—for tolerable "leakage"—a multilevel interception regime.[5] It so happens that although a very competent endo-atmospheric BMD capability could be deployed as of 1985 or 1986, an exo-atmospheric "overlay" defense, utilizing new technologies appropriate to defeat the character of offensive threat anticipated for the late 1980s and beyond, is unlikely to be ready before 1990 (although technical opinion is divided sharply on this subject). Moreover, BMD in the 1980s and beyond should be particularly effective in the context of defense of deceptively based assets—indeed, to the point where, for the offense, a catastrophic level-of-attack multiplication would be mandated for a "brute force" saturation reply. BMD of MX/MPS would not, of course, be SALT compatible, unless the terms of the ABM treaty were renegotiated—a matter of considerable interest to the Reagan administration.

To summarize, MX/MPS is not the perfect solution to the silo vulnerability problem, but it promises to be quite good enough. Much of the contemporary MX debate is being fueled by defense analysts who choose to ignore this "good enough" fact, and instead are endeavoring to press the U.S. defense community into revising the MX program in detail so that it might be marginally improved. It is very possible that the U.S. body politic, at large, fails to appreciate that much of the sound and fury that it monitors concerning the MX program consists of conflict between defense analysts who are arguing over secondary details.

WHY SHOULD THE FOLLOW-ON ICBM HAVE THE TECHNICAL CHARACTERISTICS CHOSEN FOR THE MX MISSILE?

As the late Donald Brennan asserted in talks given at Hudson Institute, [6] there is something to be said for the allegation that baseline MX is "the son of SALT." In the context of launcher limitations, it is only to be expected that the most potent ICBM payload platform would be chosen. The "base-line MX" ICBM is the highest-payload missile permissible under SALT constraints. [6] From the point of view of manpower for operation and maintenance, and efficiency more generally, it is easy to make the case for "baseline MX." However, if there were no SALT constraints on aggregate launcher numbers, or if those constraints were expressed in terms of throw weight or payload, one could well imagine the U.S. defense community deciding to spread the permitted payload among a very large number of launch vehicles. So, as a counsel of perfection, one might well prefer that the payload of 200 MX missiles be spread over, say, 1,000 launch vehicles.

The argument over more or fewer ICBMs is complicated by considerations of operation and maintenance costs and survivability. Nonetheless, overall, one cannot deny the charge that launcher limitations have tended to help drive payload to the boundary of permissibility. From the point of view of survivability and flexibility in employment, one might well prefer a proliferation of much lower payload ICBM launchers, but registration of such desiderata do not constitute proof of the folly of base-line MX design (at an 8,000-pound payload). Given the MPS and preferential active defense options available to the United States to protect a 200-strong MX ICBM force, it is difficult to make a persuasive case against the physical characteristics of base-line MX on the ground of prelaunch survivability. Postlaunch boost and very early midcourse survivability could be a far graver problem for a force of 200, as opposed to several thousand, ICBMs, if one believes that space-based high-energy-laser (HEL) BMD may be an operational reality in the 1990s, and if one believes that countermeasures to a laser ballistic missile boost-phase interception (BAMBI) system probably will be ineffective.[7] As of 1981, it is not obvious that laser BAMBI will be a reality in the 1990s, or that relatively cheap and elementary countermeasures (polishing ICBMs so as to increase reflectivity, increasing insulation, and/or the imposition of torque) or expensive countermeasures (say, involving anti-laser BAMBI satellites [HEL-equipped]) would not work.[8]

On balance, although this author recognizes the flexibility and ease of operation and maintenance of small ICBMs, and acknowledges the possibility of an adverse BMD environment wherein it would be desirable to have a very large number of launch vehicles, the dollar cost disadvantage and the farfetched character of the postulated BMD threat incline him to the view that base-line MX at worst is "good enough," and at best should be preferred over small ICBMs for the next generation of strategic missile acquisition.

Quite aside from technical issues to do with the physical dimensions of the base-line MX ICBM, there is the question of the strategic capabilities of the MX missile force considered as a whole. These two classes of issues are really quite separate, though all too often they are treated cursorily, as though they were identical. Much of the "anti-big missile" argument assumes that there is a direct connection between "bigness" and the hard-target kill capability that is deemed to be "destabilizing." In fact, the United States could pursue a truly major hard-target kill capability through development of a rather modestly dimensioned follow-on ICBM (with, say, only 2,000–3,000 pounds of payload). Such a Minuteman IV could do everything anticipated for MX, provided it were deployed in sufficient numbers. Generic critics of the base-line MX missile tend not to criticize the packing of so much hard-target lethality into one platform (a practice that merits careful examination) but, rather, the hard-target lethality itself.

Therefore, there are really two classes of dispute about the follow-on ICBM. First, defense professionals wonder whether it is wise to concentrate so much firepower on one launcher (10 x 335 kilotons)[9]—that concentration making the launcher a very high-value target indeed, prior to warhead separation. Second, there is a running argument over the desirability of procuring the capability to place at risk the more than 70 percent of its offensive-force payload that the Soviet Union has on board its ICBM force.

Certainly the United States has been driven down the MX system road far more by the determination to preserve the prelaunch survivability of an ICBM force than by a determination to increase strategic firepower. In principle, and even in some proposed practice, the missile and the basing mode issues are separable. If the United States did not wish to threaten Soviet hard targets, but did wish to retain an ICBM force, it could do the following:

Rebase a canisterized, reengineered Minuteman ICBM in an MPS system[10]

Deploy an extended-range version of the Trident (I) C-4 SLBM in an MPS system[11]

Deploy a new small or "light" (by U.S. SALT definition) ICBM in very limited numbers in an MPS system

Deploy base-line MX on a more modest scale than that proposed at present (less than 200)

Defend the existing ICBM force mix with Site Defense/LoADS BMD. (Given the scale of the Soviet offensive threat, a very large inventory of on-line ABM interceptors would be required. For warhead "leakage" to be kept low, there should be more than one "layer" to the defense. Also, this option—which certainly is a real one in terms of technical feasibility—would require renegotiation of some of the terms of the ABM treaty of 1972.)

MX missile proponents have tended to be a little shy of confronting the hard-target counterforce/alleged stability issue very directly. This bashfulness may have been politically prudent in the past, but it is probably ill-advised today. Certain facts should be widely disseminated. First, the Soviet Union has procured in its fourth (really "fourth and a half") generation ICBMs, the capability to neutralize all American hard targets. In short, the policy question today does not have the academic character of "Is a major hard-target kill capability desirable?" The Soviets have it—and seem very determined to improve it further. Indeed, it is the imminence of the maturing of this capability that is driving the U.S. ICBM modernization and survivability program.

Second, the United States has always targeted Soviet strategic forces,[12] in keeping with an evolving deterrence rationale for doing so. Secretary of Defense Harold Brown, for example, was enthusiastic about the merits of a robust second-strike counterforce capability (a counterpunch equalizer), but the only difference between this officially endorsed idea and the concept of first-strike counterforce lay in the realm of numbers. One can argue, with some good reason, that ICBMs have been rendered more lethal by what may be called "technology creep." Inexorably, weapons are "improved." American nuclear warhead designers, missile fuel technologists, and guidance engineers are professionally dedicated to "improving" their products. Although technological stagnation could occur, it is a nearly trivial point to observe that U.S. (and Soviet) high-technology industry is in the business of continuous innovation. "Stability" theorists, wisely or otherwise, cannot prohibit more efficient (energy yield to weight) weapon designs or more accurate guidance systems.

Moreover, as Jack Ruina noted about the single most important variable in missile lethality, "Guidance accuracy, there is no way to get hold of it, it is a laboratory development, and there is no way to stop progress in that field."[13] U.S. declaratory doctrine has not emphasized the benefits of hard-target counterforce since 1962–63, in substantial part because a virtue has been made of necessity. As the Soviets began placing their ICBMs in silos after 1964, and as the Yankee class SSBN became operational in 1968, U.S. counterforce prowess diminished. In the 1970s U.S. ICBM CEPs were reduced very markedly, and warhead efficiency was improved, but the deployment trend—reflected in the Poseidon C-3 SLBM, the Minuteman III ICBM, and even the Trident C-4 SLBM—was toward extensive payload fractionation (albeit MIRVed), in the context of very modestly proportioned missiles. Moreover, the Soviets superhardened (to more than 3,500 psi) many of the old SS-9, SS-11, and SS-13 silos into which they placed the fourth-generation ICBMs.[14]

Third, the U.S. doctrinal opposition of the late 1960s and 1970s to hard-target kill competence (which has been reflected not in a matching, and massive, restructuring of SIOP design, but in a growing U.S. inability to perform well the major hard-target counterforce missions that are in the SIOP) is widely recognized as almost certainly based on a fallacious mirror-image perspective on Soviet political-strategic thinking. From the early days of arms control endeavor—for example, in the context of the "Surprise Attack" conference of late 1958[15]—the U.S. defense and arms control community has worried extensively about "technical" strategic instability potential. As this author has explained in detail elsewhere, the Soviet Union does not appear to endorse Western stability theory, even implicitly.[16]

Soviet military and naval thinking is heavily imbued with the idea that the initiative should be seized at the outset of a war through a disruptive first strike. Soviet military thinkers, in line with Soviet political thinkers, appear to be almost totally unreceptive to systemic theories of conflict. The Soviets have never given evidence of believing in the possibility of war occurring through accident (fairly broadly interpreted), through the tactical mis(?)calculation reflected in the idea of "the reciprocal fear of surprise attack,"[17] or as a consequence of technical, perhaps mechanistic, instability. These Soviet data, and absence of Soviet data, are highly relevant to the MX debate, since many opponents of MX allege that that missile, deployed 200 strong, would have to promote crisis and arms race instability. So political is the Soviet perspective on conflict, that the idea that a central war could be precipitated as a consequence of the technical details of the strategic-weapon balance is close to an absurdity in the eyes of Moscow.

Stated at its baldest, the United States requires a follow-on ICBM force with the characteristics chosen for the MX because, in Soviet defense-analytic and political perspectives, such an ICBM force should be maximally deterring. In the Soviet view, deterrent effect flows from anticipated war-waging prowess. However, one does not think through what many Western theorists have deemed to be the "unthinkable" likely operational details of a nuclear campaign solely with a view to optimizing deterrence. War may occur: not by accident or because of technical crisis instabilities, as noted above—those may be shallow Western ideas—but because powers hostile to the U.S.S.R., grown desperate as the tide of history threatens to drown them, seek a way out of their problems through military adventure. The Soviet Union takes the possibility of war very seriously indeed, as it does its duties should that event occur.[18]

The case for the base-line MX missile (or, if preferred, for a much larger force of Minuteman IV missiles) may be made by direct reference to Soviet practice. An MX missile force should be both untargetable (with profit) and should threaten the survivability of the essential assets of the Soviet state. The MX ICBM, alone, cannot guarantee U.S. victory, but it should be able to guarantee Soviet defeat—a prospect that Soviet General Staff planners should find very discouraging. A survivable (50 percent) MX ICBM force would pose an unsolvable problem for Soviet strategic targeteers;[19] in addition it would place Soviet hard targets of all kinds (political, military, and economic) usefully at risk. If properly designed, a U.S. MX deployment should deter a Soviet strategic attack of any kind, and should inspire the kind of professional respect that a defense community feels for a first-class adversary.

It is important that the United States sustain a program that would eventually place Soviet ICBM payload at first- and (if held in reserve) second-strike risk. Such a capability, as reflected, for example, in a 200-strong MX force, should deter Soviet attack (by denying the Soviets a [really the] major counterforce option); should render such an attack an exercise in progressive and massively unbalanced disarmament, on the Soviet side, were it ever to be effected; and might—as a consequence of these military considerations—prompt a very large and very expensive restructuring of the Soviet strategic force posture. In addition, the silo threat is essential if the Soviets are to be denied the option of responding to U.S. ICBM modernization via MPS construction, with the simple expedient of fractionating payload through much more extensive MIRVing.

IS THE MX PROGRAM, AS CONFIGURED AT PRESENT, TECHNICALLY AND OPERATIONALLY SOUND?

Some of the engineering and operational details of a system as large and complex as MX both will and (probably) should alter over time. There is always room for improvement. The optimal way to operate and maintain MX/MPS may not become clear with respect to some details prior to the fielding of the system—there is no substitute for experience. [20] The important question at this early stage of program development relates to the basic parameters of the system: have correct, or "good enough," decisions been taken with respect to the basic elements of system design (and, hence, operation and maintenance)? For such reasons as scale of technical challenge, strategic-doctrinal ambivalence (on the part of the Carter administration), domestic politics, and arms control considerations, the MX missile and its basing mode have been subject to extraordinary vicissitudes.

Indeed, it has seemed at times as if the only constant in the MX system equation is the fact of change. Every noteworthy change in system design since 1978, even changes manifestly for the better, have done damage to the credibility of the program. The Air Force (and even the Office of the Secretary of Defense) has stalwartly defended a succession of preferred basing modes—each time with an apparent, or at least an asserted, confidence that "this time we have got it right (or right enough)." Public skepticism over the MX basing mode design of late 1980 was scarcely surprising, given the confidence with which different designs were supported in 1979 and early in 1980. [21]

The author believes that the officially preferred MX basing mode (as of late 1980) will be "good enough." Indeed, this belief was expressed admirably in The Economist (of London):

> The Carter administration has come up with yet another scheme for deploying America's planned new MX missile. Each of the past four summers has seen a different proposal for hiding and protecting this potent new MX weapon when it comes into service around 1987. Since the latest project looks more sensible than any of its predecessors, and since the MX missile represents the last chance of keeping the land-based part of America's nuclear armory safe against a Soviet surprise attack, the time seems to have come to stop arguing about the best way of deploying the MX, and actually start making the thing and putting it in the ground. [22]

The following are the principal features of the basing mode design inherited by President Reagan:

Linear-track, or grid, geometry in deployment (instead of the "loop" road announced by President Carter in September 1979).[23] This design will require less land than would the loop road scheme, and it reflects the official decision to abandon the requirement that missiles be capable of an unmanned dash to almost any shelter on receipt of tactical warning of ICBM launch. However, the linear track, or grid, in combination with the other features specified below, will permit an MX missile to be moved from a "parking" position on the road into a nearby shelter during the flight time of an SLBM.[24]

"Closed cluster" design. For SALT verifiability, as with the former "loop" arrangement, each cluster of 23 shelters along a linear track will form a cul de sac—there will be no direct connection between tracks/clusters. This arms control "requirement" (or alleged requirement) is expensive in terms of the additional road mileage that it mandates, and is undesirable from the perspective of the preservation of location uncertainty (PLU). (That is, under the proposed "closed" linear track scheme, if the Soviets locate a missile, they are, ipso facto, certain that the 22 other shelters in the cluster are empty.)[25] However, this criticism does not point to a fatal, or even very serious, flaw in the system. (This author is not friendly to the idea that the United States should offer any basing system design elements that might bear adversely upon operational PLU.) In the opinion of this author, reasonable—indeed, continuous and certain—verification standards can be met from the factory to, and out of, the main assembly areas. There is no obvious reason why each cluster should be closed. However, one should not exaggerate the benefits of cluster "connectivity." No matter what the geometry of the road layout, the Soviets know that because of technical uncertainties over the effects of closely spaced explosions, the United States would not place MX missiles in adjacent, or even nearby, shelters.

Horizontal shelters. Horizontal, as opposed to vertical, sheltering of the MX missile would permit the rapid unloading of the missile from its transporter during SLBM flight time and the fairly rapid shuffle of the system as a whole, if that were deemed advisable during a period of heightened tension. Psi resistance and cost penalties are paid for horizontal sheltering, but the former is really unimportant, while the latter appears not to be as great

as some critics of horizontal shelters have sought to argue (the
dollar cost penalty is a variable depending upon how rapid a
"shuffling" capability is desired for a vertical shelter system).
Depending upon how many transporters one chooses to buy, the
speed of ICBM/MPS shuffle could vary dramatically. However,
the important point to remember is that a vertical shelter system,
whatever else may be said for it, inherently entails a slower
"shuffle" process than does a horizontal system (raising and
lowering an ICBM—even a well-canistered ICBM—from and into
a deep hole in the ground is inherently a delicate and rather slow
operation).

A curious, almost religious, intensity developed in early
1980 over the issue of horizontal versus vertical shelters. Strong
proponents of vertical shelters argued that the nominal 600 psi
differential in favor of verticals (1,200 versus 600 psi) could well
be critical in terms of the ability of the U.S.S.R. to produce an
MX saturation capability by 1989 (the MX FOC date).[26] This
argument was supported by reference to the official U.S. national
intelligence estimate (NIE) on the U.S.S.R. "leaked" details of a
spring 1980 NIE indicated that the best estimate of Admiral Stans-
field Turner, director of the Central Intelligence Agency, was that
only "trivial" numbers of MX would survive a 1989 attack by Soviet
hard-target kill-capable RVs (with the absence of SALT limits on
Soviet warhead growth).[27] Such a Soviet saturation capability would
be achievable, so the argument went, because the relatively low
psi resistance of horizontal MX shelters would enable the Soviets
to fractionate their ICBM payload very extensively—and still achieve
high single-shot kill probabilities. This argument, though super-
ficially attractive, had a number of noteworthy fragilities.

First, the Carter administration assumed a SALT-limited
launcher constraint (that is, it demonstrated how—with U.S.
assistance [via construction of 600 psi ICBM shelters]—an allowed,
or de facto allowed, ICBM payload aggregate might be subdivided
most effectively for war-waging efficacy). However, if one can
discount the launcher limit of SALT II, what remains of the argument that 600 psi, as opposed to 1,200 psi, shelters grant the
Soviets a nearly total hard-target kill capability? If thought through,
as rarely has been the case thus far, the horizontal versus vertical
shelter issue comes down to being an argument about Soviet strategic
planning.

If, as this author believes, the Soviets have a nonnegotiable
commitment to the achievement of a war-waging/war-winning
capability, then one is obliged to believe that the Soviets will,
eventually, design the best damage-limiting offset of which they
are capable to whatever MX basing scheme the United States can

devise. To be specific, in terms of the contemporary debate, if
the United States were to endorse a vertical-shelter MX basing
scheme, then the Soviets should be expected to favor an offensive-
forces regime that would allow them to develop the payload necessary to overcome the higher psi resistance of the vertical shelter.

By way of conclusion, it is not obvious that 600 or 1,200
nominal psi resistance for MX (or other follow-on ICBM) shelters
is really of much importance. The horizontal shelter, at 600 psi
(or more), properly spaced for aim-point independence, fares
badly analytically—only if one elects to discount Soviet strategic
doctrine. When Soviet war-fighting doctrine is fully accommodated,
one realizes that the prospective psi resistance of rival shelter
designs is of very little importance. A higher nominal psi resistance,
as with vertical shelters, simply should drive the Soviets to be
unfriendly toward payload or launcher limits in the future.

Second, arguments about the competitiveness of Soviet warhead production in regard to U.S. MX shelter construction rest,
nontrivially, upon "evidence" concerning the ability of the U.S.S.R.
to produce the number and quality of nuclear warheads needed in
order to saturate an MX/MPS deployment. The U.S. intelligence
community does not have confidence in its ability to predict the
warhead production rate (at various mixes of yields) of the U.S.S.R.
through the 1980s. There is close to a 100 percent uncertainty
factor in U.S. estimates of Soviet fissile material material production, which translates, appropriately, into a large measure of
skepticism regarding "official" estimates of the threat environment
for MX/MPS.

This author is not claiming that MX/MPS could not be saturated
by Soviet offensive forces in 1988–89, only that the hard-target kill-
capable warhead inventory often credited to the Soviet Union in that
period may not, and indeed, is unlikely to happen.[28]

A "plow-out" capability. It has been determined that a canisterized
MX will be able to plow out of a horizontal shelter more cost
effectively than would be the case were it to attempt a roof break-
through. Thus far, at least, the "plow-out" scheme, entailing a
cantilever design that would be braced against the roof of the horizontal shelter, has attracted no controversy.

A "loading dock" concept. The loading dock idea, whereby an MX
missile would be loaded into, and out of, a horizontal shelter, is a
major operational refinement on the original "racetrack" concept
proposed by President Carter in September 1979. The loading dock
is responsive both to dollar cost and to operational feasibility
criticisms of the "racetrack" scheme. In September 1979 it was
proposed that an MX missile, with its integral transporter-

erector-launcher (TEL), would drive in and drive out of shelters. That idea, in principle, was admirable with respect to speed of attack-time missile movement, but it required both very large horizontal shelters and Strategic Air Command (SAC) endorsement of an operational movement idea that affronted military common sense.

The "racetrack" idea, with its integral TEL, fell foul of both very amateur and very professional opinion. Very amateur opinion found something absurd in the Buck Rogers, or perhaps (as it was called, unkindly) Rube Goldberg, idea of an unmanned dash under ICBM attack around the "racetrack." Defense professionals, in SAC and elsewhere, came to much the same conclusion. Specifically, in the context of a Soviet ICBM salvo launch, the last thing desirable would be to "open up" the MX basing system and "dash" the integral TEL from one shelter to another. Moreover, careful defense analysis showed that the MX dash capability of the "racetrack" basing mode was very likely to fail catastrophically in the event of an intelligent Soviet attack. Hence the "loading dock" concept.[29]

The MX "loading dock" preserves a useful degree of last-minute mobility—for a fairly rapid "shuffle" between shelters if there is suspicion that the deception code has been broken—but it allows for a real-time ("under attack") "dash" before an SLBM attack in a situation where the MX on its transporter is parked on the grid roadway. There is no ambition, today, of being able to dash from one shelter to virtually any other shelter.

Defense professionals disagree noticeably on the merit of the "loading dock." In the view of this author, the concept offers an acceptable compromise between very-slow-to-shuffle basing in vertical shelters and dash-under-ICBM-attack as in the original "racetrack" scheme. Specifically, some people question whether the "rapid shuffle" of the loading dock is worth the price paid in horizontal sheltering (600 as opposed to 1,200 psi resistance). Even more specifically, people ask whether it matters that the entire MX missile force could be shuffled within 12 hours (the loading dock scheme) as opposed to 48 hours (with vertical shelters).

Unfortunately for precision in defense analysis, this issue involves more than just professional strategic judgment. Quite aside from the narrow strategic arguments concerning effectiveness, the "loading dock" scheme relates very directly to the SALT aspects of MX basing and operation. The Carter administration was concerned that MX basing not entail a proliferation of what could be termed launchers (holes in the ground with ICBM support systems, or silos), and that numbers of MX missiles deployed be verifiable, from time to time, in place in their deployment areas. Vertical shelters, it has been argued, both look like silo launchers, and could defy verifiability by overhead surveillance.

It is the opinion of this author that the MX program, missile and basing mode, as proposed at present is operationally sound. However, a useful hedge against that judgment being too optimistic could, and should, be provided by means of deployment of LoADS from the outset.[30] It is important that BMD assistance, as through LoADS, should not be viewed as contributing a "final" fix for the ICBM vulnerability problem. For example, it is just possible that LoADS radar operation might compromise the PLU of the MX missile, while—as seems to be a danger at present—it is possible that the shelter spacing of a "grid" shelter deployment might be so austere as to provide extreme problems of precise aim-point prediction for the LoADS radar.[31]

6

THE THREAT

CAN MX COPE WITH EXPECTED AND GREATER-THAN-EXPECTED THREATS?

This question lies at the heart of much of the contemporary criticism of MX, from the left and the right. Unfortunately, an honest analyst has to offer two answers to the question—first, a resounding yes, and then an ambivalent maybe. Given a United States and a U.S.S.R. equally determined to compete effectively, there is no doubt that MX would win any competition against Soviet warhead growth (the basis for this claim is offered below).[1] Second, if the United States were less determined to compete than was the U.S.S.R., there would be grounds for questioning the long-term viability of MX. The case for claiming that the Soviets cannot compete successfully with MX rests upon the following considerations:

In addition to the base-line MX target set, comprising 4,600 shelters, the Soviets would have to target 350 Minuteman III silos (if the United States adhered to SALT II MIRVed platform rules), 450 Minuteman II silos, 53 Titan II silos, and 100 or so other hard U.S. military/political targets.

The Soviets would probably want to keep a noticeable fraction of their on-line ICBM force as a strategic reserve.

Base-line MX (4,6000 shelters) could be "backfilled" to 9,200 shelters (or more).

Base-line, or backfilled, MX could enjoy the services of adaptive preferential LoADS BMD.

55

Beyond LoADS, by 1990, MX survival could benefit from a BMD "overlay" (exo-atmospheric).[2]

As the MX ICBM comes on-line, from 1986 to 1989, the 70 percent-plus of Soviet offensive forces' payload residing in their ICBMs comes more and more under first-strike risk. It is reasonable to anticipate the Soviets to be willing to pay a considerable price for the invulnerability of their forces (analysts skeptical of this point are invited to offer explanations of the massive, and massively expensive, Soviet silo-hardness upgrade program of the 1970s).

Almost needless to say, although the above points are offered with high confidence as generic evidence of the fundamental robustness of MX, it cannot be denied that a pusillanimous U.S. administration could provide real-time invalidation of this line of argument. To summarize, this author is claiming that a United States manifestly willing to build base-line MX, to backfill on shelters, and to offer one—and eventually two—layer(s) of BMD, all in the context of posing a nearly total threat to the Soviet ICBM force, will hold all of the cards that really matter. A far more difficult task is the defense of MX in the context of U.S. administrations that are only lukewarm regarding the competitive resilience of the system.

It would be sensible for the United States to field an MX ICBM system that the U.S.S.R. would be unable to draw down below 50 percent in a first strike—however massive. Nonetheless, the "MX (escalation) firebreak" argument may be robust in practice, if not in force-exchange model theory, if it refers to a nominal Soviet counter-force victory that would require the allocation of close to a 20,000 RV attack in order to succeed. Pursuing the long-standing theory of complex and benign strategic-force synergism, the major upgrading of the capabilities of the U.S. air-breathing (penetrating manned bomber, and cruise missile) and SLBM forces (through provision of a hard-target kill-capable Trident II missile) could, and should, very usefully enhance the deterrent potential of a fairly robust ICBM deployment. In other words, if a follow-on U.S. ICBM deployment required, for its neutralization, close to a 20,000 Soviet RV attack (a condition easily achievable by MX system designers—with BMD assistance), the Soviets, given their strategic-cultural perspective, would be very unlikely to anticipate a minimal and measured U.S. response.[3] Thus, the deterrent merit in a particular MX basing (and supportive BMD) design could well rest critically upon the deterrent effect of the residual U.S. strategic threat—in likely Soviet estimation.

Being sensibly respectful of what is known concerning Soviet strategic judgment,[4] it is only reasonable to assume that Soviet General Staff targeteers would anticipate a U.S. presidential determination "to do his worst"—following a truly massive attack upon the CONUS-based MX ICBM force. In terms of strategic logic, a U.S. president, and Soviet defense analysts, should anticipate a paralysis of U.S. political will—after all, the United States would have essentially no ability physically to limit damage to the American homeland. However, a Soviet leadership should be expected to be very interested in the residual dyadic capabilities of the United States.

This line of argument may be dangerous, in that it could encourage the United States to adopt a somewhat relaxed stance toward the survivability of its follow-on ICBM force, while resting its real, last-line deterrent clout in the survivable "Soviet-state busting" capability of bombers, cruise missiles, and SLBMs. This author would like to remind readers that U.S. understanding of Soviet strategic targeting design is far short of perfect.[5] Moreover, if one postulates an essentially successful Soviet attack against a deceptively based follow-on U.S. ICBM, one had better, in addition, assume some noteworthy measure of enforced attenuation of U.S. air-breathing and SSBN force effectiveness.

It is unfortunate that the United States, courtesy of the Carter administration, has slowed the evolution of every major element in its strategic force posture. It is a matter of historical record that President Carter inherited an MX program with an IOC of 1983, which was slipped to July 1986 (even though the Soviet hard-target counterforce threat became more ominous after January 1977); that he aborted the B-1 penetrating manned bomber in June 1977 (which could have had an FOC in 1985 or 1986); and that he slowed the pace of Trident SSBN acquisition. The deterrent merit of an austere, deceptively based MX system has to rest substantially upon the anxieties promoted in Soviet minds by the survival prospects (conservatively assessed) of the other two legs of the strategic forces triad. Very obviously, even if the Soviets estimate that defeat of deceptively based MX (or Minuteman IV, or upgraded Trident C-4) would constitute a Pyrrhic victory in terms of payload expended, such a "victory" might still be attractive should the residual U.S. dyad fail to pose a state-defeating threat to the Soviet Union.[6]

Probably the least attractive of all strategic situations for the United States would be a context wherein an impressive, but still defeatable, MX had been purchased, while the SSBN and air-breathing strategic-force elements posed a modest—but distinctly nonfatal—threat to the prospects of Soviet war survival and recovery.

At root, the competitive resilience of deceptively based MX is an issue of presidential leadership—it is not an issue of defense analysis. The United States can maintain an ICBM force that the Soviet Union should choose not to target—which should mean that the Soviets would be unable to effect their strategic war plan. What cannot be

done is for the United States to pretend that it is arms-competitive.
A U.S. president, properly convinced of the very strong desirability
of maintaining an ICBM force, would tell the Congress and the
country that his survivability scheme had many levels of potential
policy application. President Reagan, in 1981 or 1982, would paint
the general picture of an enduring Soviet war-waging/war-winning
determination, and would defend the MX program in terms of its
ability to expand the scale of its basing mode (backfilling shelters);
adding one or two layers of BMD; and all the while MX would pose a
more and more immediate threat to Soviet silo-housed ICBM payload.
There should, and need, be no political deception about the MX
program. Considered in toto, shelters, additional shelters, LoADS
and exo-atmospheric overlay BMD, and the threat to Soviet silos
should give cardiac arrest to Soviet General Staff targeteers.

MX/MPS is not competitive with the Soviet threat only if one
assumes a United States unalerted to the danger of Soviet counterforce intentions, and unwilling to pay the modest dollar price required
to frustrate them. It is a fact that the MX ICBM, in terms of constant dollars, will be no more expensive than was the B-52 program
of the mid-1950s, while in relative perspective, strategic-weapon
expenditure is almost trivial in relation to the defense budget considered as a whole. To be specific, MX—on a high estimate—is
judged likely to cost close to $34 billion in fiscal 1980 dollars for
R, D, T, E, and procurement over nearly ten years. Assessed
rationally, by far the most important U.S. weapon system of the
1980s (the MX ICBM) is likely to cost, per annum, well under 3 percent of U.S. defense expenditures (even including a LoADS addition).

WILL THE MX PROGRAM BE TIMELY
IN RELATION TO "THE THREAT"?

The answer, unequivocally, is no. As currently planned, the
MX ICBM constitutes a 1986–89 (IOC-FOC) solution to a 1981 problem.
The U.S. policy-relevant argument in 1981 is not whether deceptively
based MX is responsive to the threat, but whether any alternative is
more relevant, and whether any necessarily (given that it is now
1981) delayed threat reaction is worth pursuing.

In the years to come, there is little doubt that the maturing of
a genuine threat to the prelaunch survivability of the U.S. silo-housed
ICBM force is going to be viewed as a cause celebre. How could the
United States have permitted the most important leg of its strategic-
forces triad to suffer from threat-compelled obsolescence? The most
relevant defenses of U.S. inaction on this front probably are to the
effect that it was hoped that the SALT process would alleviate this

problem, [7] and that Soviet ICBM CEP improved unexpectedly in 1977—78 (Soviet ICBM test firings demonstrating a CEP of close to 0.1 nautical mile). [8] Excuses such as these should carry very little weight.

It was obvious to some people from the time of the SALT I negotiations (November 1969—May 1972) that the U.S.S.R. was totally uninterested in cooperating in an endeavor to promote the dominant U.S. vision of strategic stability—while the test firings of what were to be the SS-17 and SS-19 in 1973 and 1974 proved beyond question that the U.S. silo vulnerability problem simply had not been addressed through SALT. [9]

Courtesy, inter alia, of some U.S. technology transfer in the era of detente (precision grinding tools), there was every reason to anticipate Soviet achievement of close to 0.1 nautical mile CEP on the more competent Mods of its fourth-generation ICBMs, and CEPs of less than 0.1 (say 0.07-0.08) for the fifth generation of ICBMs in the middle to late 1980s.

The fact of the matter is that survivably based MX, and all of its suggested surrogates, are at least five years late. However, the operational question is not whether MX is late, but whether it is too late. Both generic friends and generic enemies of the MX missile system have attacked, and are attacking, the program on the gound that it is not responsive in terms either of timeliness or of eventual resilience to the Soviet threat.

As may easily be surmised from the above comments, this author is highly critical of the policy-level negligence (by at least two U.S. administrations) that has permitted Minuteman vulnerability to occur on an unacceptably large scale. However, that granted, none of the currently popular alternatives to deceptively based MX offer any genuine lead-time advantages over that program. (Lest there be any misunderstanding: this author has been writing on the need to address the vulnerable-silo problem since the early 1970s, and would almost instantly "jump ship" from defending the extant MX program were he convinced that there was a "better way," as of today.)

Probably the most vociferous recent criticism of the MX program schedule (and details) has come from the conservative side of the defense policy opinion spectrum. In particular, it has been suggested that the MX lead time (July 1986 IOC) is so unresponsive to the Soviet hard-target counterforce threat that a "quick fix" is needed for the Minuteman III force. This idea is attractive, and merits very close official scrutiny. However, there are few obvious grounds for optimism regarding its technical or political-legal viability. Without defending the extant MX IOC/FOC, the following observations need to be made concerning the alleged "quick fix" that has been suggested for Minuteman:

There is probably no lead-time gain to be registered through Minuteman "fix-up" (canisterizing and reengineering) for deceptive basing, as compared with MX program acceleration. On the basis of detailed discussions with people in the missile production (as opposed to defense analysis) business, this author believes that an accelerated MX could be "on line" almost as quickly as could a "quick-fixed" Minuteman.

Minuteman "fix-up" might have to entail taking off alert status a noteworthy fraction of America's supposedly ready strategic firepower. Minuteman "quick fix" proponents have ignored the point that they propose, in effect, to draw down U.S. prompt hard-target kill capability by perhaps 10 percent during the worst period of the Soviet "threat window" as a consequence of their "quick fix" program. It is not at all obvious that the plan to solve this problem with the 140 or so "reserve" Minuteman IIIs would be adequate.

Industry personnel and Air Force officials testify, publicly and privately, that reengineering Minuteman for deceptive basing could be a technical nightmare—certainly feasible, but hardly a "quick fix."[10]

Strong advocates of "quick fixing" Minuteman have neglected to address the real pacer of their proposals, the legal process lead time pertaining to withdrawing land from currently productive farming usage and to the production of an environmental impact statement. All too often, it seems, people assume that because the Air Force has had a good experience with public reaction in the existing ICBM fields in the Great Plains, those fields could be greatly altered— for a deceptive basing mode for ICBMs—with nearly trivial political costs. There happens to be good reason to believe that such an optimistic prognosis is wrong. Moreover, and this constitutes much of the problem, it should not lightly be assumed that Minuteman II or III could be canisterized, reengineered, and then rehoused in a deceptive basing mode that could accommodate MX ICBMs when they are ready for deployment. MX, by U.S. standards, will be a very large missile with a very heavy transporter: it will require roads far harder than would be the case for a canisterized Minuteman.

MX is fair game for criticism on lead-time grounds, but any advocate of an alternative "quick-fixed" Minuteman (or Trident C-4) has an obligation to explain how the political-legal problems (delay) of land acquisition and major construction in an area of working farmland—the existing ICBM fields, by and large—would be solved. [11]

Proponents of "quick fixing" Minuteman, pending MX missile availability, tend to be less than eloquent on the subject of the
preservation of location uncertainty. The USAF's preference for
contiguous MX basing in the Great Basin of Utah-Nevada rests, in
part, upon the appreciation that that area is very sparsely popu-
lated. (Of course, the fact that there is such a low number of
inhabitants [approximately 7,200] renders the area extraordinarily
vulnerable to the negative impact of MX construction and operation.)
The far greater level of human traffic in and around the existing
Minuteman fields could present a security nightmare for the USAF
endeavoring to preserve the location uncertainty of "quick fixed"
Minuteman. [12]

Finally, the supposed "quick fixing" of Minuteman for enhanced survivability most probably would have the consequence, at the least,
of delaying the MX missile program yet further, and at most (and
perhaps more likely) of aborting the MX missile program altogether. The U.S. body politic is far more interested in preserving
an ICBM force per se than it is in redressing the payload/firepower
imbalance in relative U.S./Soviet ICBM capability.

The items cited immediately above constitute only a short list
of the doubts that this author harbors concerning the supposed promise
of "quick-fixing" Minuteman as a near-term option allegedly compatible with MX.

On the evidence available, the United States should endeavor to
accelerate the deceptively based MX program to the degree legally
and technically feasible, and attempt to fill the "bathtub" of strategic
deficiency in the early to mid-1980s with extant capability reassigned
to SIOP missions (that is, carrier aircraft), de-mothballed forces,
and a degree of bluff in strategic declaratory doctrine. The author is
not unfriendly to "quick-fixing" Minuteman, in terms of the motives
of advocates of that option. But he is both skeptical concerning the
practicality of the option and nervous lest the MX missile program be
imperiled as a consequence.

7

THE SALT CONNECTION

DOES MX REQUIRE SALT CONSTRAINTS TO BE VIABLE?

In the words of a former undersecretary of the Navy in the Carter administration:

> That was, and is, an absolutely terrible argument. Any system that depends for its viability on Soviet compliance with an agreement—an agreement that by its terms expires before the system is operational—is loony. The MX has to stand on its own feet.[1]

Proponents of deceptively based MX do not, and cannot, stand on the ground of anticipated SALT constraints as the critical basis for the viability of their preferred ICBM system. If the SALT process can confine Soviet ICBM payload fractionation and MIRV launcher numbers[2]—in a context of impressive CEP improvement (to 0.1 nautical mile or better)—then, unquestionably, that would assist base-line MX survival. However, MX proponents do not assume that the Soviet Union will lightly cooperate in aiding the survivability of MX payload. The MX ICBM, when deployed fully, could constitute a nearly total threat to a silo-housed Soviet ICBM force. Such a predictable strategic fact is so inimical to the Soviet strategic world view that one should anticipate every form of prospectively effective Soviet competition, as opposed to the cooperation that could be read into the terms of a SALT agreement.

It is certainly true to claim that the payload fractionation sublimits agreed to in SALT II negotiations were highly relevant to the prospective viability of a base-case MX deployment. However, it is no less true to claim that those sublimits would have been operationally irrelevant, since MX is not scheduled to achieve its IOC prior to July 1986. Furthermore, it is not obvious that the fractionation sublimits of SALT II constituted any restraint of military significance upon Soviet capabilities. Given their counterforce dedication, the Soviets will be interested in very extensive fractionation only when there is a strong case for it on grounds of target coverage (which will not obtain prior to the late 1980s), and when Soviet ICBM CEPs are low enough to permit a very large reduction in yield for acceptable kill probabilities. Although the Soviets have been reported as achieving test accuracies of 0.1 nautical mile, it is unlikely that most on-line SS-18s and SS-19s are assessed by Soviet defense analysts as having projected operational CEPs very close to that number.[3]

The SALT viability argument would have the U.S. body politic believe that the Soviets will agree to fractionation sublimits such that MX would retain its prelaunch survivability. Given the importance that the Soviets are known to attach to ICBMs, how likely is it that they could be brought to agree to cooperate to the effect that they would not be able to target profitably the most potent weapon system in the U.S. arsenal? Almost regardless of one's views on SALT and the desirability (or otherwise) of MX, this SALT viability thesis overstretches credulity. Since the Soviets have always approached arms control as an instrument of defense planning (when it was not purely symbolic for the encouragement of the unilateral disarmament of others), it is grossly improbable that they would compromise the military integrity of their war plans in a fundamental way, simply to preserve the SALT process. The SALT viability argument, if believed, should lead one swiftly to the position that the Soviets will not accept any SALT III agreement that constrains their ability to target deceptively based MX effectively.

There are three ways in which one might seek to find some merit in the SALT viability argument. First, one could explore the possibility that the Soviets will be so discouraged at the prospect of competing with MX (and its basing growth potential) that they will decide preemptively not to compete. The United States would, of necessity, be uncertain as to why the Soviet Union appeared to be interested in continuing (or establishing—should SALT II fail to be ratified, as seems certain today) the fractionation sublimits, and hence might be induced to pay a considerable price as (needless) compensation.

Second, one might attempt to argue that although the Soviets would be very interested in countering the MX, their genuine distaste for fractionation sublimits could be overcome through bribery, or the fear that a SALT III failure on this issue would promote a very adverse (to the Soviets) discontinuity in the level of American defense expenditure.[4] Almost anything is possible, but this author has great difficulty conceiving of any U.S. "balancing" compensation that would offset, fairly, what the Soviets (in Soviet doctrinal perspective) would judge to be the military sacrifice they would be accepting, and would be tolerable militarily and politically to the United States.[5] Although the SALT process per se does function asymmetrically to the net Soviet advantage,[6] the events of late 1979 demonstrated very clearly that in contexts where only modest advantages are anticipated to accrue from arms control, the Soviets will sacrifice arms control prospects (at least for a while).

The third, and by far the most promising, argument for the SALT dependence of MX holds that MX, when fully deployed, will render the vulnerable silo problem a bilateral one. In terms of Western strategic logic, it is not unreasonable to argue that the Soviets may choose to deny themselves the ability to overwhelm MX as the price that they must pay in order to deny the United States the reciprocal ability to acquire an RV inventory large enough to pose a very severe threat to Soviet strategic assets. It is worth noting that the United States should be able to influence Soviet decision-making with regard to the response to MX. The attractiveness of a negotiated solution to the ICBM vulnerability problem should be related rather directly to the scale of the competitive task the Soviets will anticipate should they elect to try for a brute-force saturation-capability solution. The U.S. government could strengthen the hands of those in Moscow attracted to the negotiation of fractionation sublimits in SALT III by making a credible-sounding, and credible-looking, commitment to the sustaining of the invulnerability of MX—whatever it takes (adding shelters, layers of BMD).

Notwithstanding the reasonableness of the above argument, it may be difficult for U.S. governments to appear determined enough to discourage the Soviets from selecting a brute force competitive solution to their MX problem.[7] Also, Soviet strategic culture is known to be generically unfriendly to the constraining (even the bilateral constraining) of major war-waging capabilities.[8] Moreover, the Soviets may have some good reason to believe that they would be able to maintain a survivable mobile ICBM force more easily than would the United States. They would have none of the domestic political public reaction problems that have beset U.S. ICBM basing schemes. (The Soviets need not deploy a mobile ICBM force in regions of the U.S.S.R. where local sentiment, in time of war, might constitute a severe security risk.)

Overall, the argument that MX will be viable only in the context of SALT-imposed payload fractionation sublimits is incorrect. It is correct only if one assumes an MX system frozen forever at its baseline scale of 4,600 shelters. Also, the argument is very foolish. MX should not be bought if its utility depends upon Soviet cooperation.

IS DECEPTIVELY BASED MX COMPATIBLE WITH SALT?

Before answering this question, two observations have to be made. First, for reasons quite unrelated to ICBM survivability, there may in the late 1980s, as today, be no SALT regime, or even active SALT negotiating process with which the MX system could be either compatible or incompatible.[9] Second, even if one could demonstrate incontrovertibly that MX is or would be incompatible with the SALT process in general, or with some particular SALT regime, it is not obvious that such incompatibility should be considered a major point against the system. While "arms controllability" is valid as a factor to be considered when new weapon programs are assessed, it should never be forgotten that the United States does not invest in strategic forces for their ease of arms controllability. (In a reductio ad absurdum the United States does not build strategic forces in order to limit them.)

This is not to disparage arms control per se; it is simply to assert the noncontroversial point that the MX system is designed to solve a set of very serious security problems for the United States and its allies. From extensive past experience, even directly in this area, it is axiomatic to affirm that security problems can be addressed through arms control only if they are first addressed on a unilateral basis. The United States should choose to have MX rather than SALT, if that were the choice.[10] Let it be stressed, that is not the choice—today or, prospectively, tomorrow.

There are several ways in which the MX system might be (and has been) held to be incompatible with SALT (particular agreements or the process in general). In summary form, these claimed incompatibilities lie in the realms of fundamental definition (what is an ICBM "launcher"); verification by national technical means (NTM); and competitive stimulus (with several alleged instability-promoting consequences).

The SALT II Treaty limits ICBM launchers, SLBM launchers, heavy bombers, and air-to-surface ballistic missiles "to an aggregate number not to exceed 2,400" (upon entry into force of the treaty). The SALT II Treaty does not define a "launcher"—a noticeable omission, given its centrality as a principal unit of strategic account.[11] For reasons of contemporary technical detail, and diplomatic habit, there is a SALT negotiating history that could be held to have established (through implicitly agreed usage) that an ICBM silo is an ICBM launcher. It has been true, to date, that U.S. ICBMs, to be launched, depend upon environmental controls, power supply, and communication equipment that are integral to a silo.

However, as the Air Force has demonstrated, even Minuteman II or III can be launched from a hard pad in the open. The multiple protective structures to be built for MX have little or nothing to do with the launch of the missile—they are designed exclusively to protect it from detection and destruction. As common sense, one should affirm that an MX launcher is that which launches an MX missile. The MX missile canister fulfills the reliable launch-essential functions that the silo fulfills for other ICBMs: it provides climate control, power supply, and communication access. In short, there is no technical case whatsoever for asserting that MPS proliferation would be a violation of Article III, paragraphs (1) and (2), of the SALT II Treaty. A "protective structure" is not a missile launcher.

MX/MPS deployment is verifiable by Soviet NTM, and in practice it will likely be more verifiable by NTM than is any other strategic system deployed by either side. MX/MPS probably is unique as a major weapon system, in that verifiability has been a design requirement from the very beginning. The debate has ebbed and flowed over a range of basing modes, but the requirement for verifiability has always been taken very seriously. Indeed, if one sought to assault MX/MPS on "arms control" grounds, verifiability by NTM could not, sensibly, constitute a part of the opposition case.

The official U.S. objective regarding the verifiability of a land-mobile/movable ICBM force has been to set a near-perfect precedent. Wisely, it has not been assumed that the Soviet Union necessarily would emulate U.S. MPS basing of ICBMs. The U.S. position is that the onus of verifiability falls upon the deployer, and the United States, with MX/MPS, is offering a precedent for redundant verifiability by NTM. Official U.S. reasoning holds that the Soviet Union may choose a very different (from MPS) kind of basing mode for mobile ICBMs (if they elect to go that route), but the burden of ensuring verifiability by NTM is on Soviet shoulders.[12]

This author believes that U.S. MX basing mode designers have been too accommodating in the system features intended to promote ease of verifiability by Soviet NTM. Essentially for fear of senior U.S. policymakers' disapproval, MX/MPS system designers preemptively overresponded to predictable questions regarding NTM of verification. The proper concept for MX/MPS design with reference to verifiability is the analogue of SSBN deployment. In other words, the system should be verifiable as it is deployed from the factory, or dockyard, or major maintenance facility, but not when it is "out in the field or at sea." The U.S. government has chosen to design a system that would be verifiable by NTM from the factory to the major designated assembly area, into the deployment area, and within the deployment area.

It is a matter of historical record that MX/MPS system designers were so nervous of negative arms-control-related judgments about their weapon that they deliberately grossly overdesigned MX/MPS for its verifiability by Soviet NTM. The horizontal shelters of the current basing mode were designed with four openable "ports" in the roof—for Soviet satellite inspection—and, initially, there were two "barriers" ("earth berms") an MX missile would have to traverse on its journey from the "designated assembly area" into the MPS deployment complex. In addition, the principal assembly building, outside the operational deployment area, would have a roof that could be opened for satellite inspection. Through a process of system refinement, one of the barriers has been judged superfluous, and the expensive (if not extravagant) "railroad" that was to have transported the canisterized MX missile from the designated assembly area into the deployment complex has been judged needless gilding in the verification lily. The "verification ports" in the roofs of the horizontal MPS remain, but they are certain to be removed in due course.

The proper verification requirement for MX/MPS should be "from the factory to the deployment area." Canisterized MX, like SSBNs, should disappear within its deployment area. More evidence of quite excessive zeal on the NTM of verification is provided by the fact that each MX ICBM is to be deployed within a linear "dead end" complex of MPS. The U.S. government, under no Soviet pressure whatsoever, has chosen to deny itself the capability to move MX missiles between "grid complexes." This decision adds to the cost of the basing mode and reduces the flexibility of U.S. precise deployment choices. In practice, maintenance of the "closed-grid" concept means that should the Soviet Union be able to detect a shelter that houses an MX ICBM, it will be able—courtesy of that fact—to identify 22 other shelters that do not contain an MX ICBM. Interconnectivity of 23-shelter grids would deny Soviet defense planners a certain knowledge of where MX ICBMs were not deployed. On balance, on NTM of verification issues, MX proponents need to be offensive-minded, not defensive-minded.

Without denigrating the serious and genuine verification issues that have to be (and have been) addressed with respect to MPS basing of MX, [13] it is relatively easy to lose a sense of proportion. If anticipated SALT verification problems either delay or abort an MPS program, the United States, at least tacitly, would be permitting an (and by no means the most important) arms control perspective, of a rather narrow technical kind, to control its strategic force posture. On occasion one might be willing to endorse such a situation, but—as a general condition—there can be no justification for it. The United States, or any other country that places serious security burdens on its armed forces, cannot afford to design force postures allowing somewhat technical arms control considerations to enjoy a potential veto role.

At some point in the 1980s or 1990s, the United States may have to deter a desperate Soviet leadership in a moment of very acute crisis and, possibly, may have to execute strategic nuclear employment options. In that context the absurdity of the United States having forgone, or possibly degraded much of the value of, a major strategic nuclear instrument (MX/MPS) because of (very dubious) anticipated verification difficulties, would become all too apparent. If the price that has to be paid for a very high-confidence verification capability is the design of a force posture that denies the United States necessary deterrent effect and war-fighting options, then one can only hope that a president (and Congress) would choose sensibly.

The strategic arms competition between the United States and the Soviet Union is not, at root, a cooperative enterprise. Some arms-control-oriented tracts tend to give the impression that the Soviet-American strategic relationship is a game involving cooperation and conflict in roughly equal proportions. This is not the case. The Soviets cooperate to the degree they judge (and occasionally misjudge—no omniscience in Moscow is presumed here) useful to inhibit U.S. competitive performance.[14] The Soviets have no known or logically deducible interest in designing a strategic force posture, plus complementary arms control regimes, that would effectively defang an arms competition that has evolved, and is evolving, to their net advantage. While on the one hand it is appropriate to consider the verification problems that MPS basing for U.S. ICBMs might pose for a SALT regime, on the other hand it is scarcely less appropriate to observe that the only reason why MPS basing is needed at all is that the Soviet Union has chosen to design an ICBM program with characteristics of payload, accuracy, and numbers that poses an intolerable threat to the prelaunch survivability of U.S. ICBMs housed in silos.

How serious are the verification issues associated with MX/MPS? More to the point, perhaps—and to pose a question that appears to have passed relatively underappreciated—what is the character of the American interest in this question? High-confidence monitoring (a technical intelligence function) and verification (a political function) of U.S. MX/MPS is not the problem. The Soviets know perfectly well that the United States could not cheat and hope to escape undetected. The reasons for this are diverse, but are totally persuasive. First, U.S. society, even in a heavily defense-oriented mood, is not disciplined in a way at all analogous to the Soviet Union—there would be leaks. Second, the U.S. government, and U.S. society more generally, does not subscribe in practice to the principle of caveat

emptor ("let the purchaser beware"). It is fundamentally un-
American to endorse the idea that under the terms of a contract (say,
a SALT II or SALT III), one is licensed to do anything that the other
side either does not discover, or discovers but decides not to
challenge too energetically.

As Soviet SALT practice to date has illustrated, caveat emptor
is a distressingly permanent feature of Soviet culture.[15] Thus, one
should dismiss out of hand any argument to the effect that the Soviet
Union would have genuine grounds for concern over its ability to
verify U.S. compliance with a SALT regime, should the U.S. follow
the MX/MPS path. Soviet argument to that effect would, no doubt,
meet with some sympathy in predictable quarters in the United States,
but it would not, and could not, be a serious argument.

To date, it is a matter of public record that the Soviet Union
has expressed reservations on the subject of land-mobile/movable
basing for ICBMs being compatible with the language of the SALT II
Treaty. This is hardly surprising, since the major purpose under-
pinning a U.S. move to MPS basing for ICBMs is to deny the Soviet
Union a convincing first-strike option against U.S. strategic assets.
Given the war-fighting/war-winning orientation of Soviet strategy, it
is only to be expected that Soviet SALT policy should be designed to
inhibit prospective U.S. military prowess. On the public record, at
least, there is no known Soviet position on the verification issues
pertaining to MX/MPS. In short, thus far, Americans very largely
have argued with themselves. The objections to MX/MPS that have
been devised from within the U.S. defense and arms control community (and even from outside, in the form of environmental protection considerations)[16] have been sufficiently serious, in terms of
practical politics, that one can well imagine that the Soviets believe,
or hope, that the U.S. body politic will effect the Soviet mission of
inhibiting U.S. strategic effectiveness for it.

In U.S. perspective, the MPS verification issue is potentially
serious only with respect to possible Soviet responses to the U.S.
program. It is possible, but very far from certain, that the Soviets
would "go MPS" themselves. Aside from the possibility that they
may be confident in their ability to choose the moment for strategic
nuclear initiation—and hence would feel no need for ICBM mobility—
one should contemplate the prospect of a Soviet Union that chooses a
genuinely land-mobile ICBM deployment—certainly more mobile than
are the SS-20 IRBM and the SS-16.[17] Thus, the U.S. problem may
well not be "How do we verify Soviet ICBM deployment in MPS
complexes?" MPS deployment may not appeal to the Soviet military.[18]

It is prudent to accept as facts two propositions that bear upon
the extreme ends of the Soviet missile production and deployment
system. First, it is unlikely that the United States will every be able

to monitor Soviet ICBM production with any high degree of accuracy. Second, it is very improbable that the Soviet Union will ever permit ("intrusive") on-site inspection of land-mobile ICBM complexes/ deployment areas. Proponents of U.S. MX/MPS should not rest any hopes or arguments on those two items. However, all is not lost thereby.

With respect to the confidence that the United States could and should place in adequate verification of land-mobile/movable basing for ICBMs, three elements need to be introduced. In combination they should provide all of the assurance that a nonparanoid perspective ought to require.

First, through the Standing Consultative Commission of SALT the United States negotiates with the U.S.S.R. the most detailed possible understanding concerning the monitoring of land-mobile/movable ICBMs into (and out of) deployment areas, possibly in conjunction with a random, on-site sampling system.[19] The verification challenge is highly sensitive to the precise character of the basing mode selected.[20] Since there is no solid basis, at present, for predicting which land-mobile/movable basing mode, if any, the Soviets will choose, and since the Soviets will have high confidence in American compliance with SALT provisions, for the reasons specified above, it would be absurd to consider aborting MPS basing on verification grounds.

Second, the United States should design its MPS basing structure, preferably with some BMD protection for synergistic effect (at least as an easily deployable option), so that a very considerable measure of threat insensitivity would be built in.

Third, MPS basing and associated programs should be designed so that they can expand on a discouraging (to the Soviet Union) scale in the event either of determination that the Soviet threat is far greater than had been anticipated, or of a formal Soviet "breakout" from a SALT Treaty-constrained environment.[21] The growth potential inherent in MX/MPS would help stabilize a SALT regime.

The verification problems of MX/MPS begin to fade appropriately if one takes sensible account of the promise in redundancy, rapid expansion, and a very deterring strategic nuclear employment policy.

It is easy to exaggerate the importance of land-mobile ICBM verification issues. Regardless of the character of SALT regimes, or their absence, the United States cannot monitor Soviet ICBM production; the Soviet network of major roads is not at all extensive; and any truly mobile land-based ICBM would have to have a payload considerably more modest than that of the current "heavies" (the

SS-9 and the SS-18). It is entirely possible (and indeed is probable) that if a U.S. MX/MPS deployment drove the Soviets, for survivability, to a land-mobile deployment, U.S. uncertainty over the number of Soviet ICBMs deployed would be more than compensated for by the reduced payload on the Soviet ICBM force imposed by the land-mobile deployment mode (not to mention possible additional CEP uncertainties, C^3 problems, and the increased technical downtime that would have to ensue as a consequence of frequent movement).

The third region of inquiry, arms competitive stimulus, touches upon fundamental issues of strategic perspective. In principle, there is no doubt that the prospect of MX deployment should stimulate a direct, "brute-force" Soviet response. MX, after all, is designed, inter alia, to pose a potentially near-total first-strike threat to the prelaunch survivability of Soviet ICBMs. However, generic opponents of the MX ICBM, because of its putative counterforce effectiveness, have a problem: it is Soviet hard-target counterforce programs that are driving the U.S. MX program.

It cannot be denied that MX may stimulate a Soviet strategic arms competitive response. On a similar basis, one can attempt to argue that MX/MPS deployment is unhelpful for SALT because it provides a stimulus for an offsetting Soviet deployment. Nonetheless, MX/MPS proponents should not shy away from the fact that the stimuli for their preferred system are Soviet hard-target counterforce prowess and the newly assessed need for the United States to pose a convincing threat to "the things the Soviet leaders appear to value most . . . ."[22] The U.S. government cannot afford to acquiesce in the suggestion that it should eschew the design of more survivable ICBM basing options simply because such design might stimulate the Soviet Union to prepare prospectively effective replies. In short, if taken too seriously, the "arms race stimulus" objection to MX/MPS amounts to an injunction against effective U.S. strategic arms competitive behavior.

In the early 1970s it was fairly popular to argue that arms control could prosper only in a context characterized by so-called stable arms competitive behavior.[23] In other words, negotiated limits on strategic armaments should be attractive when neither side has strong incentives to effect rapid changes in the character and/or quantity of its strategic forces. The ABM treaty of 1972 was deemed to be a major contribution to the cause of arms race stability, since it should greatly reduce the anxiety of each side over the penetration survivability of its ballistic missiles. The MX/MPS system, it could be argued, is inimical to SALT in that it provides a very large incentive to the Soviet Union to effect a massive increase in its RV inventory. The United States has been driven to MX/MPS by the pace and character of the Soviet ICBM modernization program (fully consistent with legalistic [Soviet] interpretation of the terms of SALT I). The United States cannot decline to compete effectively simply because of claimed negative impact upon SALT. First, it is not at all obvious that a robustly survivable system like MX/MPS would have a negative impact upon SALT. Second, and far more important, the fate of SALT is of distinctly secondary significance in contrast with the need for the United States to remain an effective arms competitor.

Finally, no matter how great the temptation, it is a mistake automatically to equate SALT with arms control. This author is prepared to argue that MX/MPS will serve the long-term arms control objectives of the United States, regardless of the fate of the SALT process (let alone the fate of SALT II). An arms control perspective is now incorporated into the American defense policymaking system, for good or ill. That perspective sees merit in (limited) cooperation with the adversary, and determinedly views the arms competition systematically in pursuit of the instrumental goal of stability. Today it is fashionable to refer to "the crisis of arms control" and to endorse the judgment that "arms control essentially has failed." However, on closer inspection, "the crisis" and what "essentially has failed" refer to formal interstate arms control processes as practiced from 1969 to 1979 (SALT and MBFR [Mutual and balanced for reduction negotiations])—which have borne, at best, only a tangential relationship to long-term, and long-standing, U.S. arms control objectives.[24]

To take an extreme characterization, one can conceive of an MX/MPS program that would both strain the SALT process to the limit and forward U.S. arms control objectives. Readers may recall that the two overarching arms control objectives of the United States are to reduce the risks of war occurring and to reduce the damage likely to be inflicted should war occur.[25] Historians may well come to judge that the formal institutions of arms control of the 1970s constituted a major misstep in relation to the prospects for substantive arms control achievement. The future for genuine progress in arms control is more likely to be found by reference to the subject matter of Jeremy Stone's Strategic Persuasion[26] than it is through the resurrection of formal, institutionalized SALT.

HOW WILL MX AFFECT THE SALT PROCESS?

The answer to this question, in U.S. perspective, is "positively." Soviet evaluation may be rather different. The Soviet Union, as best can be judged, has sought to employ the SALT process as an instrument of defense planning. Soviet defense planning is guided by an enduring determination to achieve such measure of military

preponderance as its U.S. competitor permits.[27] MX/MPS clearly
is extremely unwelcome, in Soviet assessment, because it is intended
both to provide a massive "escalation firebreak" and to place at
first-strike risk the 70-plus percent of Soviet offensive-force payload
that currently is housed in the silo-based ICBM force. If the Soviet
Union sees value in the SALT process in terms of the extent to which
that process discourages effective American competitive behavior,
then MX/MPS is likely to diminish Soviet interest in SALT. One is
moved to observe that a SALT process that functions to the net Soviet
advantage (because of the very different domestic contexts for defense
decision-making) is a SALT process whose demise Western democracies should not lament.

The future of the SALT process depends fundamentally upon
political, not defense-program, events. Soviet-American, and East-
West, political relations, writ large, will determine whether the
current arms control hiatus is extended or is repaired. MX/MPS may
cause problems for arms control negotiations, but it will not, alone,
cause SALT to founder.

It is entirely possible that the clear evidence of a United States
pushing forward energetically with MX/MPS would encourage the
Soviet Union to see a sharply diminishing value in SALT. After all,
in Soviet perspective, if SALT no longer can encourage delay in major
U.S. weapon programs, what value has it?[28]

In common with William Kincade of the Arms Control Associa-
tion,[29] this author suspects that the Soviet Union will be very strongly
motivated to seek a brute-force, competitive answer to MX/MPS.
However, MX/MPS, as endorsed by this author, is not a program
that Soviet defense planners should be eager to engage. A Soviet
Union motivated, for good war-fighting and war-survival reasons, to
seek a direct military counter to MX/MPS, should anticipate that
"base-line" MX/MPS (200 MX ICBMs deployed among 4,600 "protective structures") could be protected, additionally, by a doubling of
shelter construction to 9,200, provision of a LoADS, and provision of
an exo-atmospheric BMD system.

All the while, the burgeoning MX ICBM force would be posing a
more and more severe first-strike threat to Soviet silo-housed ICBM
assets. Given the technically somewhat extreme lengths to which the
Soviet Strategic Rocket Forces have gone in superhardening their
ICBM silos and launch control centers[30]—far beyond the point of
diminishing marginal return to effort[31]—it is tempting to believe
that the Soviets would be exceedingly reluctant to move substantially
to a land-mobile ICBM deployment mode. Nonetheless, in the fall of
1980 the Soviet Union did conduct static firing tests of an MX "look-
alike," one of the new fifth-generation ICBMs under development.
The very expensive superhardening process illustrates the point that

the Soviet Union is apprehensive about the prelaunch survivability of its most potent class of strategic weapon systems, the ICBM.

MX/MPS, properly designed and politically supported, could have only a benign impact upon the SALT process (in the U.S. perspective). MX/MPS would reduce the risk of war in that it would deny the Soviet Union the planning option of a devastating countermilitary first or second strike. In addition, the MX missile would make the land-based ballistic missile survivability issue bilateral. An MX/MPS program, as outlined here, would guarantee to the Soviet Union that no unilateral advantage could possibly be secured by straight, unconstrained competition. In this context Moscow would have every incentive to alleviate its strategic planning problems through substantive SALT solutions. The United States, in this scenario, would be providing the Soviet Union with a range of unpleasant alternatives. However, the fundamental point, which certainly would not be lost on the Soviet General Staff, would be that the Soviet Union enjoyed no prospect of a purely competitive success in regard to MX/MPS.

If MX/MPS were to cause the SALT process to founder, which is highly unlikely, it would be because the Soviet Union saw that weapon system as a threat it could not constrain usefully within the SALT framework, and because Soviet leaders would judge a SALT process that could not constrain MX/MPS to be a SALT process not worth preserving.

The U.S. government should not be interested in the subject of MX/MPS and arms control, per se; rather, it should be interested in MX/MPS and balanced and worthwhile measures of arms control. MX/MPS deployment can be shown to be incompatible with a SALT regime that would work in the Soviet favor. In summary form, MX/MPS could accomplish the following:

Through its separation of aim points from offensive potential, it could enable the United States to agree to a dramatic drawdown in numbers of strategic nuclear launch vehicles, while leaving considerable flexibility possible over the scale and quality of threat it was deemed desirable to pose.

Because of its hard-target kill potential and its survivability (MPS plus BMD possibilities), it would provide the Soviet Union with a very persuasive set of reasons for choosing cooperative arms race management rather than unbridled competition (that is, the principal alternative could be a U.S. hard-target threat on a possibly total scale—which could not be targeted effectively by Soviet forces).

It would be compatible with a militarily (and hence politically) intelligent strategic nuclear employment policy.

As an ongoing program, it would constitute a warm production line
that could become hot if the need arose. For stability in an arms
control regime, there have to be credible and prospectively effective hedges against "breakout" events.

It is always tempting, for the sake of simplicity and manage-
ability of analysis, to isolate the variable of principal concern. In
this context it is worth noting that MX/MPS, however the program
progresses or fails to progress, is unlikely to play the lead in any
scenario involving the definitive collapse of the SALT institution.
SALT II should have been defeated for its inherent lack of strategic
merit, but the events of the fall and winter of 1979 demonstrated, yet
again, that the fortunes of arms control agreements can have relatively little to do with their strategic quality. [32]
Virtually two generations of American arms controllers (the
first generation, naturally enough, has trained its successor via
Ph.D. dissertation supervision and influence over major foundation
grants) have encouraged, and still encourage, acceptance of a set of
ideas that diminishes the likelihood that its bearers will appropriately
appreciate the potential SALT payoff, for international security, of a
serious MX/MPS program. Without denying the merit, a priori, in
the charges that MX/MPS could pose crisis and arms race instability,
and SALT verification problems, it is reasonable to observe that a
very different perspective on MX/MPS is no less worthy of adoption.
To be specific, MX/MPS can be defended as being the U.S. strategic
weapon program that speaks most directly to the Soviet perspective
on central war. In a well-designed MX/MPS, the United States would
have an instrument capable of prompt hard-target counterforce kill,
able to deter attack upon itself, and very capable of eliminating a
large fraction of the withheld portion of the Soviet ICBM force. [33]
Instead of focusing upon the problems that MX/MPS might pose for
the SALT process (let us wait for the Soviets to do that), why not ask
what MX/MPS might accomplish for the SALT process ?
Notwithstanding the new realism with which the U.S. defense
community has come to view the Soviet military buildup (very largely
since the Team "B" exercise on the national intelligence estimates
regarding Soviet intentions, conducted in 1976), [34] the degree to
which MX/MPS proponents have argued defensively is noteworthy.
The reasons for that defensiveness are not difficult to locate: they
include suspicion that President Carter's commitment to the program
may have been highly contingent upon temporary political considerations; realization that the missile's characteristics represent something of a historic turnaround when translated into doctrinal and
war-planning terms—notwithstanding the unveiling of some details
concerning Secretary Brown's "countervailing strategy"; [35] sensitivities

over possible or alleged adverse environmental impacts; embarrassingly high financial costs; and (last and definitely least)
putative SALT verification problems.

What is very noticeable is that MX/MPS proponents have taken
due—even undue—notice of every actual and potential criticism that
has been, or might be, leveled at the system. But opponents of
MX/MPS, at least to date, have not reciprocated by taking proper
notice of the arguments favorable to MX/MPS. There is a danger
that a subcommunity favoring development of a particular weapon
becomes so tightly locked into a defensive mode of argument that it
neglects to make a convincing positive case for its preferred system.
It is not enough simply to suggest, even very plausibly, why the major
charges leveled against MX/MPS are without merit; skeptics need to
be persuaded of what MX/MPS could accomplish for Western security.
A negative argument rebutted is not synonymous with a positive argument advanced.

The arms control and strategic necessity cases for MX/MPS
cannot sensibly be disentangled from one another. In very general
terms, the United States cannot (save by luck) have an arms control
policy that has integrity unless it first has a coherent national strategy
(declaratory policy, weapons procured and under development, and
detailed war plans). Whatever the tactical negotiating benefit of MX
may be in the SALT context, the United States needs a robust prior
understanding of which future postures are consistent with doctrinal
desiderata and which are not. Before looking at the SALT negotiating
connection of an ongoing MX/MPS program, several general strategic
points need to be made.

First, a timely MX/MPS program (timely, that is, in relation
to prudent, though not necessarily "worst case," estimates of the
evolving threat to silo-housed Minuteman) should greatly reduce the
perceived U.S. need for SALT-imposed ceilings on "the threat."
Second, an MX program of appropriate scale would enable the United
States to execute the "countervailing strategy" that has been proclaimed. Third, MX/MPS has the virtue, particularly if supported by
a healthy preferential BMD deployment option waiting "in the wings,"
of imposing an all but insurmountable escalation control barrier:
confronted with MX/MPS (and perhaps some BMD), the Soviet Union
could not execute a large-scale successful counterforce strike. The
importance of this point would be difficult to exaggerate, given the
known counterforce proclivity of official Soviet military thinking.
This author has yet to see a generically anti-MX/MPS analysis that
took due and fair account of these arguments. On the political front
it would be difficult to improve on the cautious language of Harold
Brown:

Given the past importance of our ICBM force and the
traditional emphasis of the Soviets (and of many military
observers throughout the world) on ICBMs, it can be
argued that a decision not to modernize the ICBM force
would be perceived by the Soviets, and perhaps by
others, as demonstrating U.S. willingness to accept
inferiority, or at least as evidence that we were not
competitive in a major (indeed, what the Soviets have
chosen as the major) area of strategic power.[36]

The above perceptual theme of argument is inherently a "soft"
one, but—that admitted—it has to be difficult to deny that for the
United States to fail to modernize its land-based missile force, or to
phase it out altogether, would constitute a very unhelpful backdrop
to U.S. (SALT) diplomacy in the 1980s.

The arms control negotiating case of MX/MPS is very straightforward. First, MX/MPS would compel Soviet leaders to consider
seriously arguments in favor of a major drawdown in land-based
missile forces. (Of course, other alternatives would be available:
they could vary their strategic forces mix more in favor of sea- or
air-based systems, or they could elect to invest in a land-mobile/
movable system of their own. However, MX/MPS should present
them with a real decision point concerning the future structure of their
strategic forces.) If SALT is to be about disarmament, rather than
simply the licensing of extant programs, the Soviet Union has to be
provided with a major incentive to disarm by mutual agreement—
MX/MPS prospectively is the most plausible and persuasive incentive
toward such an end.

Second, MX/MPS is required in order to maintain U.S. confidence in the SALT process. Because of its "firebreak" potential to
thwart Soviet counterforce war planning, MX should greatly ease
U.S. anxieties over the risks being run through acceptance of SALT
constraints. A U.S. land-based missile force that could not be
targeted efficiently, and that constituted the keystone in SIOP execution, should significantly diminish potential American anxieties over
the prowess of SA-10s and SA-11s against LRCMs and B-52s, and
should usefully redirect some Soviet resources away from ASW research and operations.

Third, with MX/MPS the United States could afford to be fairly
relaxed over the pace and detailed content of SALT negotiations.
MX/MPS, with a BMD hedge for synergisticic protection, affords a
relatively SALT-insensitive strategic posture. Needless to say,
U.S. SALT negotiators should be able to perform more effectively
than in the past, if they are not constantly "watching the clock" to
seek to induce the Soviet Union to sign an agreement that would

largely alleviate U.S. strategic problems before intolerable threats matured.[37] Experience in all fields of negotiation suggests that the side in most need of an agreement is the side likely to secure the least advantageous bargaining outcome.[38]

Fourth, an MX/MPS system (with or without BMD assistance) has an inherent flexibility of scale that should appeal to all sides in the U.S. defense and arms control debate. MPS could accommodate a severely constrained threat potential (such as deployment of the Trident C-4 missile), or it could accommodate "base-case" MX with ten reentry vehicles. The counterforce threat in this system is separate from the basing mode—meaning that aim points ("protective structures") and independently targetable warheads need bear no numerical relation to one another. An MX/MPS deployment of very modest size, say 150–200 missiles in 23-strong MPS complexes, could be expanded fairly rapidly should the need arise. It is true, as some analysts have suggested, that MPS basing concentrates costs (of construction, largely) in the near term, while benefits flow much later.

If this is judged to be a telling argument against MPS basing, it reflects a sad commentary upon the quality of recent U.S. defense policy. How can it be that a country as wealthy as the United States might reject a superior weapon system, on the ground that it would generate a useful level of military muscle perhaps one to two years later than some of its (otherwise inferior) competitors? In fact, this theme of argument has no merit whatsoever. The Carter administration imposed delay after delay on the MX program and its basing mode, thereby implying that a year here or there in the initiation of its substantial contribution to strategic deterrence should not be judged a matter of serious concern.[39] (This author did not endorse the relaxed timetable of the Carter administration, but he is convinced that an IOC of meaningful proportions—say, approximately 1,000 protective structures—could be achieved one year earlier than seems probable on the existing schedule).

Fifth, for arms control to "succeed," in American terms, the Soviet Union has to be convinced that the alternatives to balanced and substantive agreements are less desirable than are negotiated constraints. As Harold Brown implied, in the words quoted earlier, an ICBM modernization program by the United States tells the Soviet Union, in a language that it has no difficulty comprehending, that America is determined to compete effectively. As a backdrop to the SALT process, how would it be if, during the lifespan of a SALT II, the United States chose to move to a strategic forces dyad, because of the Soviet hard-target counterforce threat, and then—in the course of a SALT III regime—elected to move to a monad of SSBNs as a consequence of the increasingly severe Soviet air defense menace to

U.S. LRCMs and (non)penetrating manned bombers? It is essential that the Soviet Union respect the United States as a worthy adversary. Unilateral U.S. restructuring from a triad to a dyad (or less) would invite Soviet contempt.

No one can, or should, claim that an MX/MPS program assuredly will lead to a SALT regime that will satisfy many American arms control ambitions. The outer limit of the argument in this study is to the effect that with MX/MPS a satisfactory SALT agreement may be possible, and that should SALT negotiations break down, MX/MPS would help enable the United States and its allies to endure a period of possibly enhanced competition with considerable confidence (although MX/MPS is not a panacea for the political and military problems of the Western alliance).

8

ALTERNATIVES TO MX AND
THE PROBLEMS OF LEAD TIME

Even though MX/MPS details keep evolving, there is a strong case to be made that the U.S. ICBM survivability problem has, if anything, been overstudied.[1] It is worth recalling Henry Kissinger's judgment that "further study" is a standard bureaucratic ploy intended to produce delay: "Research often becomes a means to buy time and to assuage consciences. Studying a problem can turn into an escape from coming to grips with it."[2]

Some aspects of the MX/MPS system will benefit constantly, albeit perhaps trivially, from further study, if only because the political and strategic environments are dynamic. Almost needless to say, the greater the delay imposed by study (inter alia) schedules, the more likely it is that substantial elements of the system in question will need to be changed.

MX/MPS has been, and continues to be, the target of what warrants description as a "technological filibuster." The debate over MX/MPS is difficult to close so long as rival system concepts constantly are advanced. The latest in a long series of supposed alternatives is the Submersible Underwater Missile System (or SUMS)—previously known as the Shallow Underwater Missile System. The new title (keeping the old acronym), with its ridiculous redundancy—submersible and underwater—had to be coined in order to take account of the potentially lethal Van Doren (a tidal wave) effect in shallow water. The principal function, objectively speaking, of SUMS and the like is to keep alive the debate over the proper solution to the vulnerable silo problem. Because the Soviet military threat is constantly evolving, there can be no definitive solutions to defense problems.

Some of the critics of MX/MPS who have taken to citing allegedly unmanageable offensive threats to the system[3] are wont to forget that

they themselves must bear part of the blame for the delay in the
MX/MPS program that has helped fuel the high threat-level anxieties.
The United States cannot be defended with studies. With respect to
ICBMs, as to other kinds of military equipment, a point comes
where the government should decide that a given configuration is
"good enough." The United States cannot enjoy the benefits of
security in the long term unless it survives the near term. The delay
of three years in MX IOC imposed by the Carter administration—for
a mix of reasons—could translate into Western political and/or
military defeat in the mid-to-late 1980s. MX is not "just another"
weapon system.[4] President Reagan may be able to reduce MX/MPS
lead time by one or two years (particularly the lead time to FOC),
but that can be achieved only through adoption of a new weapon
acquisition philosophy and through a willingness to bear the political
costs of short-circuiting the excessively long land acquisition and
environmental processes that currently have the force of law.

Persons other than defense professionals may be attracted to
the argument for further study of alternatives because of the apparent complexity of the MX/MPS system, its cost, or its possible
impact upon arms control processes. Such people may be unaware
that complex mobility schemes for ICBMs have been studied for more
than 20 years. Indeed, some noticeable fraction of the study effort of
recent years has probably been unnecessary—given the library of
such studies already on government shelves. The government is
desperately short of genuinely strategic studies of the need for MX/
MPS, but then—as this author has argued many times[5]—the market
for strategy in Washington tends to be episodic and small.

In summary:

The alternative of phasing out ICBMs altogether, or of choosing to
live with a silo-housed ICBM force that would have value only in
a first-strike role, has been judged unacceptable by virtually all
respected commentators.

The alternative of adopting, operationally, a policy of launch under
attack is judged to be irresponsible and possibly very dangerous.[6]

MX/MPS, generically, is superior on strategic, political, economic,
and lead time grounds to its principal competitors (air-mobile
ICBMs, air-transportable land-and-launch ICBMs, SUMS).

MX/MPS has enough growth potential (as a target set) in several
respects (shelter increase, BMD synergism), to be plainly "the
best buy" for the mid-to-late 1980s through the 1990s.

This author has developed critiques of alternatives to MX/MPS elsewhere,[7] and so will not repeat the details here, save in a summary fashion. The principal idea currently in some vogue as an alternative to MX/MPS is SUMS, which envisages deployment of a fleet (more than 100) of small, diesel-powered submarines/submersibles that would carry three or four canisterized ballistic missiles on their hulls. Richard Garwin and Sidney Drell, the principal advocates of SUMS, have claimed that this system would be more timely, cheaper, and more survivable than deceptively based MX.[8] Predictably, this program has attracted some favorable commentary (after all, is it not obvious that the U.S. government would reject anything that was relatively cheap, timely, and survivable?).[9] As it happened, the Carter Department of Defense proved to be vulnerable to SUMS argumentation both because the extant MX program reflected a relaxed, peacetime, "business as usual" mentality, and because it chose to attack SUMS on an ill-judged basis.

To be specific on the latter point, the Department of Defense elected to destroy SUMS by reference to the "Van Doren effect" of nuclear explosions in shallow water. Government witnesses, including the undersecretary of defense for research and development, William Perry, told Congress that SUMS submarines could be destroyed easily by the massive shock waves and turbulence that would be caused by nuclear explosions in and above the waters of the continental shelf (the Van Doren effect of shock waves off a shallow ocean floor). While Perry was probably correct, he neglected to notice that SUMS submarines could operate off the west coast of the United States, where there is virtually no continental shelf, or in deeper water in the Atlantic beyond the continental shelf. On a more serious level, SUMS is not a candidate for official U.S. adoption for the following reasons:

Its lead time would be longer than that of MX.

Its dollar cost, considered in appropriate system perspective, could could easily be double that of MX.

It offers no performance advantages over the Trident program.

It would require a massive increase in the ASW capabilities of the U.S. Navy. The defense of 100-plus SUMS submarines would be a major new mission.

Not to mince words, SUMS is not a serious suggestion. This author, who is critical of the MX lead time of the Carter administration, can see no major advantage of any kind for SUMS. SUMS was invented by people who knew little about designing submarines or operating navies.[10]

One somewhat popular alternative scheme to MX/MPS, air-mobile ICBMs (either released and fired in midair, or transported to randomly selected ground-based launch sites), [11] has a fundamental strategic flaw: it would mean that two legs of the strategic forces triad would be dependent upon warning for "dash mobility" survival. The choice of a short takeoff and landing (STOL) aircraft as the ICBM carrier would minimize the threat to prelaunch survivability, but air mobility would constitute a rejection of the basic reasoning that underlies the triad. That is, each leg of the strategic posture should offer a distinctive target structure, compelling a very substantial dispersal of Soviet effort.

Aside from survivability concerns, an air-mobile and air-launched ICBM would have serious operational implications. The accuracy of such a system would be inferior to that attainable from fixed sites on the ground, and C^3 similarly would be more tenuous. Also, the acquisition costs for the fleet of specialized ICBM carriers, and the operation and maintenance costs of the total system, would have to be prodigious. [12] In its "land and launch" version an air-transportable MX ICBM would have the same undesirable features, in orthodox stability terms, as would MX/MPS, plus a few additional ones. Specifically and preeminently, a "base-line MX" ICBM, or any variant close to it in payload and accuracy, would (if deployed in sufficient numbers) pose a very large threat to Soviet strategic capability—heavily concentrated as it is in superhard silos—but this would be an American threat that the Soviets could neutralize if they struck fast enough. Unlike MX/MPS, a "land and launch" MX could not absorb punishment—if its carrier aircraft could not escape safely from runways, the system would be neutralized. This system looked more like a lightning rod for a surprise attack by SLBMs fired on depressed trajectories, than a stabilizing successor to silo-housed Minuteman.

Without doubt the Soviet Union would have a very difficult time orchestrating with high confidence a total suppressive attack against "land and launch" ICBMs, but the idea has such an obvious weakness (apart from dollar cost) in its dependence on warning time for the safe escape of aircraft, that it should be judged inherently inferior to concepts that are not dependent upon warning time for survival.

It is probably true to say that the most likely alternative to MX/MPS is precisely nothing (or perhaps some small augmentation in the scale of the airborne and maritime legs of the consequent strategic dyad). [13] As recently as 1978–79, President Carter was reported to be unpersuaded of the merits of retaining an ICBM force of any character in the United States. [14] To be fair, Mr. Carter almost certainly would have preferred to negotiate the demise of the ICBM force, rather than to abandon it unilaterally without obvious reciprocal sacrifice on the Soviet part, and he did endorse an ICBM modernization program. However, his fundamental skepticism over the value of a strategic triad that included a large force of ICBMs was common knowledge.

Some elements in the Carter administration certainly favored a more effective countermilitary nuclear targeting capability, but it is worth remembering that PD 59 was in need of implementation. If, as reported, the United States plans to lay heavier targeting emphasis upon Soviet projection forces, the readiness, accuracy, flexibility in retargeting, short flight time, and exceptionally secure C^3 (as compared with submarines or aircraft) of ICBMs render a land-based ICBM force a necessity (not just a desirably diversifying element in a strategic forces mix). If one asks what the costs might be of shifting the structure of the strategic forces to a dyad excluding ICBMs, the considerations discussed immediately below should weigh heavily in the balance.

Careful and rational defense analysis plays only a limited role in political perceptions. Even if it could be demonstrated analytically that the United States would be well advised to move to a strategic forces dyad comprising SLBMs and bombers/cruise missile carriers (N.B.: it cannot be so demonstrated, plausibly), the cost in loss of international political reputation could be very noticeable. Like it or not, ICBMs are perceptually prominent. Few politicians or officials in any country may comprehend many of the technical and strategic arguments pertaining to the desirability of ICBMs, but a unilateral, and undeniably coerced, American evacuation of its ICBM silos would not pass unnoticed. [15]

American achievement of its goals in SALT III and beyond could be compromised fatally if the ICBM force is not modernized. Strategic-force reduction on a nonmarginal scale is dear to Mr. Reagan's (as it was to Mr. Carter's) heart, and is explicitly written into the statement of principles for the guidance of SALT III negotiations, which is an integral part of the SALT II package. [16] Rightly or wrongly, the American arms control community believes that SSBNs/SLBMs and manned aircraft/cruise missiles are inherently stabilizing elements in a strategic force posture, and that any relative shift of inventory emphasis to those elements, away from land-based missiles, is a gain for stability. [17]

Accepting this logic for the limited purpose of this discussion (and only for this restricted purpose), it is fairly obvious that MX/MPS is a critically important element in any theory that purports to explain why the Soviet Union might agree to shift the extremely heavy emphasis it has placed upon land-based strategic missiles toward the more stabilizing SLBM and airborne elements. If superhard Soviet ICBM silos are not credibly threatened by an American MX ICBM program, where is the profit for the U.S.S.R. in reducing its ICBM deployment on a substantial scale? It is true, in theory, that the much-delayed hard-target counterforce potential of American ALCMs could be impressive, but there are some practical details that serve to discount the political benefit of that threat.

Above all else, a Soviet Union liberated from the need to plan a saturation attack against American ICBM complexes would be a Soviet Union liberated to devote truly massive resources to the neutralization and attrition of the American dyad—and particularly of the delayed hard-target counterforce potential of that dyad. [18] It cannot be asserted, with very high confidence, that MX/MPS must promote real disarmament (of the ICBM forces) in SALT: the Soviets, in principle, do have the alternative of a competitive response to MX/MPS, but there can never be guarantees attached to strategic predictions. With MX/MPS there is some prospect of the U.S.S.R.'s negotiating a SALT agreement that would draw down ICBM inventories on a major scale, while without MX/MPS it would lack the incentive to redirect its strategic deployment program in so fundamental a way.

MPS basing for American ICBMs might have the dominant consequence of fueling Soviet ICBM payload fractionation (to the limit licensed by SALT II) and, generally, of driving the Soviet Union to the development of a strategic saturation targeting potential. No advocate of MX/MPS should deny this possibility. But, as already noted, there are cost-effective ways of discouraging this Soviet response and, even if this response cannot be deterred, MX/MPS still should be the preferred American ICBM modernization option.

In relation to SALT, MX/MPS embraces a range of acceptable risk. At worst, MX/MPS drives the Soviet Union into a cost-ineffective competition that it loses; at best, it compels the Soviet Union to agree to a drastic drawdown in ICBM force levels.

Next, MX/MPS will have technical characteristics that a U.S. president would value particularly highly in the event that prewar deterrence broke down. MX/MPS will enable a president to draw down the residual throw weight in the Soviet ICBM force to a noteworthy degree, [19] and will offer that flexibility, accuracy, and promptness of response that a serious war-fighting strategy requires. In terms of deterrence, MX/MPS would be a strategic capability that, in and of itself, would function as an escalation firebreak—it would be a capability that the Soviet Union could not target with advantage. It is sensible to question the value of an ICBM force per se—one can only hope that President Reagan understands why a strategic force posture confined to aircraft and cruise missiles and to SLBMs would severely restrict his operational choice in a period of acute crisis. SLBMs have accuracy, C^3, flexibility, and readiness problems, while aircraft and cruise missiles would be slow to arrive on their targets (meaning, in some cases, that the target might have moved out of the lethal radius) and would be subject to attrition by the defense.[20]

The above items represent only a very few reasons why a president should not abandon an ICBM capability. An alternative to MPS ICBM basing that attracts irregular attention is the tactical expedient of launch on warning (LOW) or, for the more popular variant, launch under attack (LUA). LOW/LUA has been debated in great detail since the late 1960s; indeed, the definitive critique of the idea appeared in that period.[21] Far from being a cheap and simple "fix" to the problem of silo vulnerability, LOW/LUA is an accident-prone tactic of dubious technical feasibility that is virtually devoid of operational strategic merit.

To be specific, LUA is as much technical aspiration as a likely feasible reality.[22] The United States could expect "warning," of sorts, through reception and analysis of launch-signature information via early-warning satellite-based sensors, but "assessment" is something else entirely. In practice, at least at the present time, this author predicts that the U.S. defense community would not know, in real-time terms, whether it had to launch in response to a substantial, a very substantial, or a potentially overwhelming scale of Soviet missile attack:[23] that is, we probably could launch, but we probably could not launch under assessment (worthy of the qualification).

If the U.S. defense community should decide that its preferred solution to the prelaunch vulnerability of Minuteman is LOW/LUA,[24] then it had better face the prospect that, in all likelihood, it would have to launch ignorant of the character of an attack. It is true that technical developments will soon improve U.S. attack assessment capability, but the fact will remain that it is basically undesirable to have the fate of 100 million (plus) Americans resting upon the quality of judgment exercised, and advice received, by a president in the space of only a few minutes. In pursuit of greater wisdom, even King Solomon may have preferred to sleep on some of his more fateful decisions.[25]

SHOULD MX/MPS BE DELAYED, PENDING THE ACCUMULATION OF MORE EVIDENCE CONCERNING THE EVOLUTION OF THE SOVIET THREAT?

It appears that some critics of MX/MPS are almost hoping that the Soviet Union will provide military program evidence of a determination to attempt to procure a successful MX/MPS saturation capability. As Paul Davis has pointed out, some people like to argue both that silo-housed Minuteman missiles are not really vulnerable, because of the real-world operational problems facing Soviet planners, and that, should the United States foolishly purchase MX/MPS, the Soviet Union could easily deploy a convincing saturation attack capability.[26] In other words, if a successfully coordinated Soviet attack against 1,054 silos is judged to be strategic fiction, in good part for technical reasons, should we reject MPS basing on the ground that a coordinated Soviet attack against 4,600-plus MX shelters will be technically feasible by the late 1980s? Davis is correct in labeling such a complex position as hypocritical.

MX/MPS proponents, while granting that the scale of the threat anticipated is indeed relevant to the U.S. debate, believe both that MX/MPS is convincingly competitive even against "superthreats" and that Soviet strategic program design and execution is, to some admittedly questionable degree, contingent upon at least the broad parameters of U.S. weapon programs.[27]

In answer to this question, there is nothing to wait for. Further delay in the MX/MPS program, unless adequate short-term compensation is provided, will simply prolong the period of American strategic inferiority. What kind of evidence might one have in mind?

Evidence of a much more cooperative Soviet approach to "stability" questions in SALT, perhaps? Unfortunately, it is too late. MPS basing addresses a problem that is already with us. MX/MPS was badly needed in the context of SALT II; now that SALT II is defunct, there is even less basis for urging program delay pending arms control action on the vulnerable silo problem (unless one believes that MX/MPS is competition-resilient only in the context of SALT constraints).

Evidence of a Soviet determination and ability to overwhelm MX/MPS, perhaps? It should never be forgotten that the superpowers are competing, and that that competition could be a matter of life or death. A number of Western commentators are all too easily deterred from endorsing what is needed (in this case, MX/MPS) by the thought that the Soviet Union will oppose it. One should not neglect the fact that the Soviet Union also faces problems of competitive interaction. As chess players, Soviet officials will know that a United States prepared to spend $30—35 billion for base-line MX/MPS would be a United States very likely willing to invest billions more should a convincing-looking threat emerge. MX/MPS, overseen by a nondefeatist U.S. government, would offer a target set predictably beyond the competitive reach of the Soviet Union. Technically unsound pessimistic commentary on the competition resilience of MX/MPS may serve the Soviet interest very well.

The folly of delaying MX/MPS further, pending the arrival of better threat data, may be demonstrated most easily by posing the obvious follow-on question, "How long should the United States wait?" The unacceptably vulnerable condition of silo-housed Minuteman is a fact today. The United States needs the capability of the MX ICBM now. Although the Soviet Union can drive up the cost to the United States of achieving and sustaining that capability, it cannot—given the program "hedges" available—deny the United States an ICBM force viable in second-strike roles save at prohibitive cost (and possibly not even then). This prohibitive cost could include the need to deploy well in excess of 30,000 ICBM warheads (to target up to 9,200 MPSs, two on one, and to effect total destruction of a BMD system that may deploy more than one interceptor to defend each MX linear-grid complex). These 30,000 warheads, let it be added, would be "exchanged" for only 200 MX ICBMs (2,000 warheads); even for a truly dedicated counterforce planner, that is an extravagantly unfavorable ratio.

Given the need of U.S. targeting strategy for the promptly available and accurate firepower of the MX, the United States cannot afford to wait, supinely, for the provision of evidence regarding Soviet competitive intentions. Although there are good reasons to believe that the Soviet Union will choose not to engage in a reentry vehicle versus MX shelter competition, this judgment may be wrong, and Soviet leaders certainly would have a strong motive for attempting to persuade the American defense community that it was their intention to compete. From Moscow's perspective, it must be comforting to hear American voices predicting the very early obsolescence of baseline MX/MPS—as Soviet ICBM payload fractionation proceeds apace (naked of SALT constraint) in the mid-1980s. The moral of this story, of course, is that one should not compete, or appear to compete, with a truly serious arms race competitor like the Soviet Union, in a halfhearted way.[28]

HOW LONG WILL MX BE A USEFUL STRATEGIC WEAPON?

There is no definitive, magic date when a system loses all utility for the national defense, and still less can one predict such a date well in advance. Weapons are procured for many reasons—some having to do with threat perception, some with the believed "ripeness of technology,"[29] some with reference to foreign and/or domestic politics, and some with reference to the technical inefficiency of attempting to maintain "old" weapons. Very few weapons actually "wear out" in any close to literal sense. This is particularly true of ballistic missiles, which spend their active lives suspended in climate-controlled silos.

However, it has certainly been the case that major missile systems such as the three Minuteman generations were not designed to last for much longer than ten years: solid fuel loses its reliability after a time, as do some kinds of fissile material. In short, aging missiles require refueling (which is not an easy task with solid fuel), may require new warheads, and will require spare parts that are no longer being manufactured.[30] As a matter of historical record, it is worth noting that, regardless of the length of active service life originally anticipated, Minuteman I was operational for 12 years (1962-74); Minuteman II, to date, has been operational for 15 years (1966-81); and Titan II has seen service for 19 years (1962—81). If these are precedents, MX/MPS might be expected to be in service from 1986 until perhaps 2001—06.

Leaving aside the issues of technical reliability (such as how reliable the MX's fuel and warheads will be by 2000 or so), it is extraordinarily difficult to decide on strategic grounds that a weapon system has become obsolete. After all, at least until now, U.S. strategic weapons have never had a realistic field test. On strategic grounds one would judge MX no longer to be useful if and when it could not survive a surprise attack, and/or it could not execute its offensive missions. However, short of a massive Soviet attack on MX/MPS, and/or actual duels between MX payload and Soviet ballistic missile defenses, one could not be certain that the system was, or was not, obsolete. A very similar problem pertains to assessment of the prelaunch and postlaunch survivability of B-52s both as penetrating bombers and as cruise missile carriers. Very firmly held opinions exist on both sides of the argument—but both sides lack evidence.

In contrast with the 1950s and 1960s, MX system designers would be prudent to assume a somewhat dilatory U.S. research, development, and procurement cycle. It is a matter of record that major strategic missile systems, thus far, have not been designed to enjoy more than ten years of operational life. It is often the case that a weapon system enjoys an honorable, active, old age—despite facts of obsolescence and increasing technical unreliability. For example, the B-52 (last produced in 1962) will still be flying in the 1990s; the FOC—currently 151—of ALCM-equipped B-52Gs was scheduled by the Carter administration to be attained in 1990, though President Reagan may succeed in accelerating that conversion program by five years.[31] Also, there are no plans at present to retire the single-warhead Titan IIs (54) or Minuteman IIs (450). The fact that a weapon system represents 20 (or more)-year-old technology does not, ipso facto, mean that it can play no useful strategic role. Nonetheless, the inventory management philosophy of "never throw anything away" has been far more characteristic of the U.S.S.R. than it has of the United States.[32]

It should be remembered that missiles, like aircraft, ships, and tanks, are weapon platforms: midcareer, or even late career major refit is far from unusual. Of the 550 Minuteman IIIs deployed, for example, 300 currently are being modernized with the Mk12A warhead, which offers almost twice the yield (for the same size and weight) as the Mk12. It is possible that the MX ICBM might, in the late 1980s or in the 1990s, have to be fitted with maneuvering re-entry vehicles (MaRVs), should the need arise to be able to penetrate a substantial conventional (not directed-energy) ABM system.

Although the pace of dramatic military technological change has slowed markedly since the 1950s, MX system designers and Defense Department officials have to consider strategic/technical threats. that could unpleasantly shorten the service life of the system. This author, for example, has considered both "conventional" threats (growth in the Soviet warhead count and Soviet deployment of many more ABM interceptors) and "unconventional" threats (directed-energy weapons). Because the United States could add shelters and/or missiles to the base-line system, add one or more layers of BMD to that system, and procure advanced penetration aids, the conventional threats appear to be distinctly manageable. The unconventional threats to MX almost defy prudent provision. It has to be admitted that there is an outside chance that the Soviet Union might deploy in space a weaponized laser or particle beam BMD system that would end the strategic authority of ballistic missiles. However, at the present time the U.S. technical community is sharply divided over the strategic value even of the nearer-term unconventional technology—that of lasers for BMD purposes. [33]

The prudent attitude toward directed-energy BMD threats to MX would seem to be one of careful attention, not panic. It would be very foolish to choose to forgo a strategic capability (MX) that is desperately needed for deterrence and war-fighting purposes just because it might be catastrophically vulnerable at some point in the 1990s (or later).

On balance, it is the judgment of this author that the space-based directed-energy BMD threat to ballistic missiles will become a reality, but not on a time scale relevant to MX planners today. (It is probably worth adding that MX may be particularly vulnerable to directed-energy BMD systems both because of its physical size and because of the relatively small numbers that are to be procured.) As a more general thought, readers should be reminded of the fact that there are counters, of greater or lesser efficiency, to most weapon systems. One does not forgo a capability, or declare it obso-lete, simply because a counterweapon exists. Counterweapons, in their turn, often have to cope with counter-counterweapons, while those counterweapons, to have a decisive effect, have to be deployed in sufficient numbers in the right places and at the right times (also, they have to work well enough under varied conditions).[34]

There is no final or absolute weapon; certainly the proponents of the MX ICBM make no such claim for their preferred system. That system is preferred for the 1980s and the 1990s. That is all.

9

MX/MPS AND THE ARMS RACE

WILL MX/MPS STIMULATE THE ARMS RACE?

Si vis pacem, para bellum neatly encapsulates a fundamental truth for the guidance of the defense programs of democracies (authoritarian regimes tend not to need the reminder). This axiom applies to the answer that must be given to this question. The essential answer has two parts and a major caveat. First, it is far more important that the West deter war adequately than that it engage in unilateral moves intended to damp the arms race—to the limited degree that those policy themes are likely to be in conflict. Second, the arms competitive response that is MX/MPS (it is not an arms competitive initiative in the way that MIRV was, for example) is fully as likely to have a medium-to-long-term damping impact upon the arms competition as to have a stimulating effect. The caveat is that this author—indeed, the U.S. defense community at large—is not privy to the intimate details of Soviet strategic nuclear-force planning and, in particular, to the details of the explicitly contingent competitive elements in that planning (as opposed to the observable and predictable, year-in and year-out, process of Soviet postural modernization).

However, the Soviet response to MX/MPS (addressed in detail in this chapter) is unlikely to be either a totally "wild card"—beyond the realm of intelligent and informed Western speculation—or uninfluenced by the way in which the United States goes about its MX/MPS business. Overall, if the United States is truly serious about procuring a survivable MX ICBM force, it would be very well advised to convey persuasively to the Soviet Union the depth of that seriousness. Somewhat lukewarm endorsement of MX/MPS might well fall short of conveying serious intent to the Soviet Union. Liberal ideology tells us that arms races are "bad" because, through mechanisms that have yet to be explored convincingly, they tend to promote wars.[1] This ideology, or theory, notwithstanding its poverty (it does not have great explanatory authority), has always been popular on the left. It acquired numerous adherents in the center of the political spectrum during the late 1960s and the early 1970s, both as a response to the militarism that was perceived in U.S. policy in Vietnam (and then, by extension, in all of U.S. "cold war" foreign policy) and in response to the accumulation of new strategic weapons in an era of proclaimed "sufficiency."

Liberal-center tolerance of "the arms race as usual" probably was markedly reduced by the way in which the then secretary of defense, Robert McNamara, chose to discuss U.S. nuclear strategy. Explaining the mid-to-late-1960s, Henry Rowen claimed:

> The primary purpose of the Assured Destruction capabilities doctrine was to provide a metric for deciding how much force was enough: it provided a basis for denying service and Congressional claims for more money for strategic forces. It also served the purpose of dramatizing for the Congress and the public the awful consequences of large-scale nuclear war and its inappropriateness as an instrument of policy.[2] (Emphasis added.)

McNamara succeeded well, and almost certainly too well, in discouraging whatever optimism over nuclear use might have been lurking in the body politic. Indeed, as Rowen went on to say:

> Increasingly what was communicated to the American people, the Europeans, and the Russians, was the prospect of 100 million dead Americans and a similar number of dead Russians (and also of dead Europeans) if a nuclear exchange were to occur.[3]

It was difficult to promote ideas of nuclear strategy in such a climate.[4] If it is a technological fact that nuclear war comes only in the 100-million-deaths-a-side variety, and that there can be no efficient defense against ballistic missiles, then why waste billions of dollars on "improved" nuclear arms, or assume the putative risks of arms competition? As President Eisenhower once observed, laconically, "Enough is certainly a'plenty."

The argument that MX/MPS will stimulate the arms race has nowhere near the political force that the identical argument had when it was advanced in 1970 in opposition to the Safeguard ABM system. What has changed? In contrast with 1970, the following are true:

There is a virtual consensus in U.S. defense policymaking and
commenting circles that the Soviet Union does not endorse the
idea of mutual assured destruction as a strategic desideratum
promoting "stability."

Although there is considerable disagreement over the state of the
strategic balance today and the political (and perhaps military)
consequences of a continuation of the trend of recent years, there
is near unanimity on the following: the strategic trend through the
1970s was unfavorable to the United States; a continuation of that
trend is not prudent; and the United States has to attempt to step
beyond the hothouse of its own strategic culture and consider its
strategic competitive needs in the light of the apparent motives,
and the manifest program actions, of its arms race competitor.

There is a rapidly growing belief, across the political spectrum of
U.S. defense opinion, that a large measure of what has been
termed "war-fighting" capability may be essential for an adequate
deterrent posture.[5]

In 1970 it was fashionable to argue that the principal security
problem facing the United States was the arms race itself—and that
the arms race, to a very large degree, was the continuing product of
a U.S. national security bureaucracy incapable of recognizing
sufficiency, and of military and industrial organizations that had very
strong vested interests in opposing any very rigorous definition of
such a concept. This thesis translated roughly into the proposition
that "the enemy" is really the way in which U.S. society organizes
for, and manages, its defense function. Hence, arms control should
be effected "at home."

The spate of articles, theses, and books on "bureaucratic
politics" and the weapon procurement process duly reflected this
inward turn to commentary.[6] This is not to deny that many of those
works were of value;[7] rather, it is only to suggest that bureaucratic
and other domestic process models of arms race behavior, while
healthily enriching understanding, could not stand alone as a promising
basis for theories of arms race dynamics. This author contributed
modestly to the inward-looking early 1970s literature,[8] so he feels
licensed to offer some measured criticism of a trend that was healthy
only up to a point.

Domestic process models of arms race behavior tend, naturally
enough, to have grave difficulty accommodating the overarching
framework of interstate competition. Rich studies of the U.S. policymaking process can be enlightening, but if one is interested in arms
race phenomena, there have to be scarcely less rich case studies of
Soviet policymaking. Progress has been recorded in understanding

Soviet style in defense matters, but the data base for Western theorists remains perilously thin.

Even if this author is incorrect in believing that the arms race should not be reified as a danger to mankind, the case for MX/MPS is not weakened by answers to this question. One may choose to allow that there are indeed two principal dangers: that the United States may stimulate the arms race and that the United States may risk prolonging military inferiority by choosing to forgo MX/MPS. It is not at all obvious that the choice is this stark—just as it is not at all obvious that stimulating the arms race need necessarily have any very undesirable consequences. If taken to the point of logical absurdity, there is an arms race stimulation case to be made against nearly every Western defensive measure. To condemn MX/MPS on the anticipation of stimulation of the arms race would be analogous to blaming a defender for choosing to fight an invader—thereby starting a war.[9]

It is reasonably obvious that throughout the 1970s the Soviet Union was not moved in its military policy consideration by any of the mutual stimulation models of arms race dynamics that have been popular in the West. For whatever blend of reasons, the U.S.S.R. has chosen to modernize and augment its strategic forces in such a way that the security interests of other countries require an across-the-board military response. So strong is the trend in the Soviet favor that arguments concerning arms race stimulation, even if plausible, are swamped by the seriousness of the military-political arguments for response. The United States cannot afford an official theory of arms race dynamics that has the effect of precluding timely military responses. The consequences of an MX-stimulated arms race _may_ be unpleasant; the consequences of a Soviet strategic modernization program that is left improperly answered certainly _will_ be unpleasant (and could prove fatal). There really is no choice.

Arms race stability is a concept derived, in its current form, from the era of mutual assured destruction in U.S. strategic thinking. It was believed, and is still believed by many, that the superpower arms competition could be stabilized by the mutual acquisition of assured destruction capabilities. Indeed, in the mid-to-late 1960s senior U.S. officials stated very explicitly that they believed Soviet strategic policy to be comprehensible in familiar U.S. terms. For example, Alain Enthoven said:

> I believe that part of the Soviets' strategic policy is the
> maintenance of an assured destruction capability. The
> Soviets clearly have the means to maintain this capability.
> Thus, the Soviets can and would react to any steps we
> might take to achieve a full first-strike capability or to
> limit damage to ourselves.[10]

The logic was very simple, and even compelling. Each side would be content only with a secure second-strike retaliatory capability of awesome potential. Any time this capability was threatened by offensive or defensive developments on the other side, offsetting action would be undertaken. Therefore, programs intended to limit damage (hard-target counterforce, ABM, civil defense, strategic ASW) would fuel the arms race through the threat that they would pose to the assured destruction capability of the other side. It was recognized that "normal" modernization and replacement of the arsenal inevitably would create some instability, but it was believed that the fueling of first-strike anxieties could be minimized through the exercise of self-restraint (for instance, in 1969, President Nixon eschewed much of the light urban-area coverage potential of the Sentinel ABM system in favor of a hard-site defense orientation, renamed Safeguard) and the negotiation of arms control agreements that would foreclose, or slow down, strategic developments that should, on this theory, spur countervailing moves abroad.

At a less specific level, "arms race stability" tends also to be employed to refer to the pace of quantitative and qualitative change in competitive armaments. This usage is vague, since it neglects to consider the fact that a fairly rapid pace of qualitative change in weaponry need not be inconsistent with a stability defined in terms of strategic accomplishment and political perceptions.[11] Indeed, in a path-breaking essay on arms race theory, Samuel Huntington argued that there is an inherent tendency toward equality of accomplishment, with its attendant "stability," in a qualitative (as opposed to a quantitative) competition.[12]

One of the great unexamined premises of modern arms control theory and practice is the idea that it is desirable to constrain high technology (through tacit or explicit agreement). This discussion will not endorse the idea that the arms competition is better left untouched by arms control attention, but it does register the thought that the idea is by no means ridiculous. After all, institutionalized arms control is an instrumental activity, of no merit in and of itself. If the ends of arms control can be forwarded more effectively by other means, then the case for formal arms control collapses. It so happens that the arms control case for MX/MPS embraces both SALT and SALTless futures. Even if the current hiatus in SALT were to become a permanent breakdown, MX/MPS could still have a salutary arms control impact.[13]

The arms race stability case against MX/MPS, as observed above, amounts to the claim that this system will drive the Soviet Union to invest massively in an offsetting deployment that, in its net effect, will leave the United States no more secure than it was at the outset of MX/MPS deployment, spurring the United States to deploy offsets to the Soviet offsets . . . and so on—the classic "mindless momentum of the arms race."

The most obvious deficiency in the arms race instability theory is the fact that nobody, inside or outside government, has yet designed a convincing model of the dynamics of the Soviet-American arms competition.[14] So, while this author, in common with others, is willing to speculate as to the likely character, if any, of a Soviet response to MX/MPS, he—again, in common with others—has no means of predicting the Soviet response with high confidence. As observed earlier in this discussion, there is enormous inertia in the Soviet weapon procurement cycle, and the sensitivity of Soviet weapon design and deployment patterns to developments abroad can easily be exaggerated. The Soviets have accumulated an investment of daunting size in their very hard silo complexes for ICBMs: early and wide-spread evacuation of those silos in favor of mobile or "movable" ICBM deployment would not be a decision taken lightly or rapidly.

Two Soviet tendencies, which perhaps merit description as traditions, suggest alternative kinds of Soviet reactions to MX/MPS. On the one hand, Soviet military posture and practice are designed to do sensible military things (to the best of our knowledge, the Soviets do not intend to employ their strategic forces in pursuit of political bargaining advantage in a process of competitive escalation, for example). Hence, if the Soviets discern the evolution of a major threat to their land-based missile assets, one would anticipate their undertaking militarily rational offsetting programs. On the other hand, the inertia in the Soviet system promotes a tendency to stability in programs, almost regardless of objective changes in the strategic environment. The philosophies of particular weapon design bureaus, in conjunction with stable doctrine on the part of military user organizations, married to the inertia that permeates the entire system, argue against abrupt changes in the course and character of weapon development and deployment.[15]

Proponents of MX/MPS should not be driven, defensively, to deny all possible validity to the arms race instability thesis. A fundamental argument, which needs to be advanced without apology, is to the effect that instability in the arms race is not the worst of the possible strategic conditions with which the United States may have to contend. Arms race instability of a very serious kind is not proved, probable, or even very plausible with regard to MX/MPS; but even if that were not the case, and MX/MPS had to be judged guilty of fueling the arms race, the following argument would have to be considered: there are no stabilizing alternatives to MX/MPS development and deployment. This is because:

American transition to a dyadic strategic posture would very
   probably be both strategically and politically destabilizing.

American adoption of some variant of LOW/LUA could be intensely
   destabilizing in time of crisis.

American deployment of preferential hard-site BMD, although
   favored strongly by this author (in synergistic conjunction with
   MX/MPS), would require substantial renegotiation (if not outright
   abrogation) of the ABM treaty of 1972, and would reopen many of
   the old arguments over BMD's potential for undermining the
   whole basis of mutual deterrence.

   So, even if MX/MPS were assessed as very likely to contribute
to arms race instability, it is important to recognize that the United
States has permitted itself to enter a period in which there are no
options with regard to the modernization of its strategic force
posture that are, a priori, innocent on standard stability criteria.
However, this category of objection to MX/MPS is vulnerable to
challenge on two generic grounds. First, it implicitly denies that
important qualities of stability can obtain in a period of rapid force
modernization. Second, it elevates a subordinate (stability—by
orthodox and highly challengeable definition) over a superordinate
concern (providing adequately for the national defense). It cannot
be repeated too often that it is the evolution of Soviet weapon programs that is driving the arms competition today.
   It is possible, and by no means wildly improbable, that the
Soviet military establishment may be far more relaxed about the
hard-target counterforce threat posed by MX ICBM deployment than
most Western analysts predict. Soviet strategic thinking has never
endorsed many of the injunctions that the American defense community believed flow from proper appreciation of the first strike/
second strike distinction. As best we can tell, the Soviet Union does
not rigidly endorse the proposition that preemption is preferable to
retaliation (indeed, the superhardening of late-model Soviet silos,
and the modernization of older ones, indicate flexibility, or perhaps
just prudent reasoning), but the preemptive theme—to go first in the
last resort—is strong in Soviet military literature.
   Since Albert Wohlstetter and his associates at RAND produced
their justly celebrated "bases study" in the early 1950s, [16] the U.S.
strategic posture has been designed and operated with a view to its
minimum vulnerability, even on a day-to-day, nongenerated alert
basis, to surprise attacks "out of the blue." The Soviet Union has
operated its strategic force posture in a noticeably different fashion.
For a combination of technical and political judgmental reasons,
Soviet strategic nuclear forces typically have been maintained on a

peacetime alert status quite dramatically less onerous than have the strategic forces of the United States. The Soviet Union has behaved, in regard to the "readiness" of its strategic forces, as if it would be able to choose the timetable for war. A country confident that it will be able to strike first need not burden its military machine with high readiness requirements on a day-by-day basis. Times have changed since the late 1950s and early 1960s, a period when the Soviet Union could in theory have been the victim of a forcibly disarming first strike, but the eventual mutual vulnerability of hard targets (assuming U.S. purchase of the MX ICBM) is likely to be viewed with less alarm in the U.S.S.R. than in the United States. For the Soviet Union vulnerability is a familiar condition, and one that should induce only modest anxiety if there is confidence that the military initiative could be seized at the outset.

Behind this particular part of the discussion is the fact that the Soviet Union has shown no willingness to seek joint arms control assistance for the timely alleviation of the vulnerable silo problem. It is unclear whether this attitude reflects the judgment that vulnerable silos are, first and foremost, an American problem—and therefore a problem that the Soviet Union has a positive interest in not helping to solve—or whether Soviet strategic doctrine and operating practices accord this issue far less importance than has been (and remains) the case in the United States.

Frequently one hears the argument that MX/MPS proponents fail to take adequate account of the competitive options open to the Soviet Union. It has been suggested that the Soviets might be able to defeat MX/MPS through very extensive fractionation of missile payload—a fractionation that Representative Robert Carr claimed would be cheaper to effect than would be the extension of the MPS system by the United States;[17] the Soviets might "go MPS" themselves, which would provide them with a hard basing structure into which secretly produced missiles could be introduced in a hurry, providing—even without very extensive fractionation—an impressive capability against MX/MPS; and that the Soviets could stockpile the necessary MPS-defeating missile force (when added to the missiles deployed in the field) and launch it from the factory or warehouse. An argument in the background of these claims is that the relaxed schedule for MX/MPS, with a full operating capability of the system unlikely much before 1989–90, would provide the Soviet Union with ample lead time to design and execute an appropriately countervailing response. (It should be noted that it is difficult to advance this argument in the same breath that one deplores MX/MPS for its alleged contribution to crisis instability.)

Clearly, MX's contribution to stability is facilitated if meaningful payload fractionation limits hold through 1985 and beyond (or if

severe throw weight restrictions are imposed), and if its basing
mode can absorb the damage that might be wrought by "a greater
than observed threat"—that is, if the United States built a noteworthy measure of redundancy into the MPS system. Just how much
redundancy would be built in is a function of the uncertainty over
Soviet ICBM production rates, possible Soviet covert payload fractionation, and Soviet targeting rules. For reasons of technical
unreliability, inadequate single-shot kill probability (or, perhaps,
lack of confidence in test-derived CEP estimates), and a prudent
concern to be really certain that the hard-target counterforce job is
"done right" (the decision to launch a full-scale attack upon the U.S.
ICBM force would have to be judged the most important decision
ever taken by any group of Soviet leaders), it is plausible to argue
that the Soviets have a targeting rule of directing two cross-targeted
warheads against every hardened U.S. military target. [18]

The case for MX/MPS does not stand or fall on the ability of
the Soviet Union to defeat the system. Instead, the case includes the
following elements: the Soviet Union could not defeat MPS basing
save at a cost in allocated payload that should be prohibitive; such a
defeat of MX/MPS would be theoretically achievable only as a consequence of an attack of gigantic proportions; and, while the Soviets
might harbor the aspiration that an attack on the Minuteman fields
only would meet with only a limited U.S. reply, they could hardly
sustain such an aspiration in the context of an attack on MX/MPS
that involved 10,000–20,000 warheads (or more). In short, MX/MPS
dramatically raises the threshold for counterforce adventure.

WHAT WILL BE THE LIKELY EFFECT OF MX/MPS
ON SOVIET PLANS AND PROGRAMS?

The Soviet Union has a tradition of pursuing cost-ineffective
missions—of doing the best that it is able, even in pursuit of the
impossible. [19] This has tended to be the case particularly when the
policy in question involved a quantitative assault on a problem, and
when changes in policy course would have had a widespread ripple
effect throughout Soviet defense industry and would have affected the
political relationship between rival military bureaucracies. In other
words, thus author believes that a well-designed base-line MX/MPS
system could have a healthy and nearly traumatic impact upon Soviet
strategic planning. But Soviet military planners may not respond to
MX/MPS as Western rational defense analysis suggests they should.
Properly designed, and sustained, MX/MPS should mean the following:

The U.S.S.R. cannot attack the U.S. ICBM force with any hope of profit (a catastrophic development for a country that has a thoroughgoing "war-fighting" doctrine).

The silo-housed ICBM assets of the U.S.S.R. would be at first-strike risk.

All fixed hardened sites would become vulnerable (always provided the United States purchased sufficient MX warheads, that the warheads could penetrate deep underground, and that the fixed targets were of known location).

Part of the beauty of the MX/MPS concept is that the hard-target counterforce threat posed by the MX ICBMs should serve very usefully to constrain the range of intelligent Soviet responses. A Soviet planner has to worry about negating the war-fighting (deterrent—in Soviet perspective) prowess of MX through Soviet offensive action, and about the possibility of the counter-MX design's incorporating some very unwelcome features. For example, some simpleminded commentators in the West imagine that the Soviet Union will be able to fractionate the payload on its SS-18 force (from the present, SALT II-authorized 10 to 20–30)—leading eventually to an SS-18 force load totaling 6,000–9,000 warheads. This would be a formidable opening bid for the discouragement of U.S. MX/MPS managers. However, such prognoses of extreme fractionation neglect to recognize that those 6,000–9,000 hypothetical warheads would be housed on board only (roughly) 300 SS-18 ICBMs. No prizes in mathematics will be awarded for the appreciation that the possible cost-exchange ratio for 300-600 well-placed warheads would be impressive.

Like a supertanker, military policy in the Soviet Union does not lend itself to agile navigation. Defense analysts in the West increasingly have come to the opinion that although Soviet military policy is not totally impervious to outside influence, there is a degree of eigendynamik, or self-directedness, that is massively resistant to external pressure. It is unclear exactly which sporting simile most nearly conveys the character of Soviet-American arms competition, but it is abundantly clear that any simile pointing to rapid initiative and response is wide of the mark. An economy substantially directed in its priority allocation policy by five-year plans is not the most agile of instruments with which to conduct a military competition.[20]

The least painful Soviet response to MX/MPS would be simply to augment current programs. Given that Soviet ICBM doctrine would appear to be a development of artillery thinking, MX/MPS invites an augmentation of counterbattery fire—more ammunition.[21] Warhead fractionation should proceed, as CEP and efficiency in warhead design permit, and a very straightforward bid for a saturation potential would be pursued. The unfortunate vulnerability of such an emerging saturation capability to a first strike by a fraction of the U.S. MX force may be discounted on political grounds (the United States would not strike first—given the near absence of a U.S. civil defense program), [22] or ignored on the basis of high confidence that the U.S.S.R. could, and would, always launch on warning or under attack. Neither of these arguments is very persuasive.

It is very unlikely that Soviet defense planners rate as negligible the possibility of a U.S. massive counterforce first strike. History and Soviet doctrine (carefully based upon historical study) tell the military scholar that the surprise attacker enjoys the great advantage of the initiative. Soviet military planners, we may be sure, would prefer that the U.S.S.R. strike first—so, in best mirror-image fashion, what could be more reasonable than to project that preference upon the adversary? Also, it is easy to forget that Soviet officials are obliged, by their legitimizing ideology, to define the United States as a deadly enemy. Again through projection, Soviet military planners are very likely to anticipate that American officials categorize them in a similar fashion. Finally, it is a fact that the Soviet Union has superhardened many of its ICBM silos and ICBM launch control centers—hardly the (very expensive) action of a state supremely confident that it will never be compelled to strike second. [23]

With these kinds of plausible Soviet considerations in mind, the following are the principal choices that MX/MPS poses for the Soviet military establishment:

Retain a silo deployment mode for ICBMs, and assume that Soviet
    ICBMs will never have to strike second
Add land-mobile ICBMs to the existing silo-housed ICBM structure
Substitute land-mobile ICBMs for the silo-housed ICBM force
Abandon ICBMs altogether—in favor of SLBMs, cruise missiles, and
    penetrating manned bombers
Abandon most of the silo-housed ICBM force, leaving enough for
    strategically interesting first-strike tasks
Retain ICBMs (MPS or silo-housed) and defend with BMD
Seek some negotiated solution via SALT.

Of these seven options, two may be discarded at the outset as being extraordinarily improbable—they affront too many major tenets of Soviet strategic culture. It is extremely unlikely that the U.S.S.R. would decide to abandon ICBMs altogether: such a decision would affront the artillery mentality that continues to dominate Soviet strategic thinking, just as it would affront the "rational" war-waging theme that has been central to the development of Soviet strategic programs.

Next, it is entirely improbable that the Soviet Union will seek to neutralize prospective MX prowess through SALT. Even if the U.S.S.R. would like to effect a general retreat from hardened-silo killing ability, it is too late. Both superpowers have demonstrated test CEPs in the region of 0.1 nautical mile (roughly 600 feet); and both superpowers have mastered the technology of MIRVing ICBM payload. However, the major test of standard U.S. strategic stability theory reposes in the Soviet threat posed to MX/MPS. To date there is no programmatic evidence that would suggest a Soviet determination to overwhelm MX/MPS. But Soviet adherence to a war-waging/war-winning doctrine apparently has been unshaken by the prospect of MX/MPS, so it is only prudent to assume that the full scope of the Soviet challenge to MX will emerge once the American dedication to the system is secured, literally, in concrete. [24]

If the Soviet Union chooses to pursue the warhead multiplication path to threaten MX/MPS, it should not be forgotten that such an endeavor could offer more and more lucrative targets to a U.S. first strike. The prompt hard-target counterforce capability of MX denies the Soviet Union the sensible option of multiplying the number of warheads already deployed on the MIRVed fourth generation of ICBMs housed in silos vulnerable to MX.

It cannot be denied that U.S. deployment of MX/MPS would confront the U.S.S.R. with an unwanted and unpleasant choice. Inalienable Soviet war-fighting doctrine requires the seeking of direct ways to counter the MX/MPS, but the only really promising route for damage limitation lies in BMD and civil defense.

Relatively few American commentators seem to understand that the United States has no prudent strategic choice other than to invest in a follow-on survivable ICBM system. To concede to the Soviets, unanswered, an ICBM capability is to concede strategic superiority. [25] This author cannot predict in detail how the Soviet Union will respond to MX/MPS, but he can stipulate the kind of Soviet threats that MX/MPS must be capable of thwarting. It is plausible to assume that the Soviets will not lightly abandon the mission of comprehensive hard-target counterforce, nor will they sit idly by, watching their increasingly fractionated ICBM payload become more and more vulnerable to a first strike by MX. Similarly, though this is particularly difficult to accommodate with any precision, it should never be forgotten that the Soviet Union does not have infinite resources for the solution of only one defense task (the negation of MX/MPS), no matter how important.

American MX/MPS planners would do well to consider that by 1989–90 the following may be possible:

The Soviet hard-target kill-capable ICBM warhead count might be
   somewhere in the range of 10,000–20,000, with 10,000–15,000
   being the more likely subrange, because of warhead production
   bottleneck problems and competitive discouragement.

A growing fraction of Soviet ICBMs might be fairly free-ranging and
   land-mobile, as opposed to "moveable," like MX. There will be
   substantial uncertainties over Soviet ICBM deployment, not merely
   over Soviet ICBM production (as today).

   It is possible that neither base-line MX/MPS (with 4,600
shelters) nor even "backfilled" MX/MPS (with 9,200 shelters) may be
able to contain the growth of the Soviet hard-target threat. Strictly
speaking, a 9,200-shelter MX/MPS deployment should defeat even a
"bad" 1989-90 case (assuming, perhaps too optimistically, two-on-
one targeting, and substantial Soviet ICBM allocation for other [than
MX/MPS] hard targets and for reserve force duties), but this author
chooses to assume that a doubling of the MPS numbers would likely
invoke very considerable political criticism from the deployment areas.
   Clearly, it is the job of the MX/MPS planner to present Soviet
General Staff targeteers with an impossible (not just a difficult) task.
This can be done. A combination of some "backfill" of shelters, beyond
the 4,600 base-line figure, and BMD should offer total discouragement
to Soviet war planners. MX/MPS with "backfill" and LoADS (not to
mention a BMD overlay) should defeat even the highest end of the
National Intelligence Estimate range of guesses regarding the Soviet
ICBM warhead count for 1989–90 (and that is a very high end indeed).
   It is unlikely, though certainly not impossible, that the United
States will have to invest in MPS "backfill" and in multitier BMD in
order to contain the Soviet ICBM threat. However, it is only prudent
to assume that the Soviets may seek to discourage U.S. investment in
MX/MPS through signaling a future ICBM warhead count that could
lie in the 10,000–20,000 range by the end of the 1980s.
   The facts of the matter are agreeably elementary. It is a
question of simple arithmetic and political determination. The Soviet
Union cannot overwhelm an MX/MPS system that is substantially
"backfilled" and is defended actively by LoADS and, if need be, by
an exo-atmospheric "overlay." If the U.S. defense community would
reflect this fact in its statements on MX/MPS, as it should, then there
is every likelihood that the U.S.S.R. would seek some nonmilitary-
competitive path for the negation of the MX ICBM system.

10

CRISIS STABILITY

In a series of brilliantly argued papers in the late 1950s and early 1960s, Thomas Schelling articulated the dangers that would, or could, attend a strategic weapons context characterized by "the reciprocal fear of surprise attack."[1] He, and others, proceeded to outline ideas for the mutual stabilization of the strategic weapons balance. The fundamental danger, as outlined at the time, was analogous to that of two gunfighters in the old West: assuming tolerably equal competence of aim, he who drew first, won. If two (theoretically) vulnerable strategic force postures faced off in a crisis, the incentive to launch first (for victory) could be well nigh irresistible.

Many defense analysts were impressed by the precedent of 1914 and its preceding decade. At that time it was very widely accepted as true that mobilization meant war (one could not mobilize and "hold"), and that the side that completed its mobilization first would win.[2] In other words, military dynamics would take over from real-time political decision (this would be Clausewitz's "grammar" of war, which in this case would have dominated political logic).[3] In the early to mid-1960s the United States moved as rapidly as possible to reduce the vulnerability of its strategic forces (total dependence on manned aircraft [unsheltered] and very soft-sited Atlas and Titan ICBMs), and espoused the theory that the goal of strategic stability would be forwarded very markedly by a similar Soviet move to invulnerable basing for its strategic forces.

Strange to say, there appears to have been little debate in the U.S. defense and arms control community in the mid-1960s over the possible (or, as this author would argue, inherent) tension between strategic stability defined as the mutual ability to inflict unacceptable punishment by way of retaliation, and the backstop role of U.S.

strategic forces in the context of possible theater crises. Even if there is a genuine stability in the exchange of such threats, is this the kind of stability that should operate to the U.S. advantage? The U.S. strategic force posture has extended deterrent duties, a fact that implies a willingness to initiate strategic nuclear employment. A secure second-strike capability to devastate 200-plus Soviet cities should (on this theory of deterrence, at least) deter the Soviet Union from initiating a large attack against American cities, and should remove any acute pressure upon the Soviet leadership to preempt.

But the stability in the mutual countervalue threat should put a lid on the military dynamics of the war—in all probability leaving the Soviet Union in possession of those Western assets that the strategic employment process was initiated to defend. Actually to devastate Soviet cities would, of course, neither defend western Europe, nor provide the Soviet Union with much of an incentive to exercise restraint. In short, the crisis stability theory of the 1960s was grossly negligent in the context of a U.S. president's likely operational needs. What the United States needed, on this theory, was a force posture that would not induce extreme trigger-happiness in nervous Politburo members in time of crisis, yet would enable a president to execute a politically sensible strategy in time of war.

Crisis stability theory in the MAD genre, whatever its merits as a theory for war prevention, failed to take account of the richness of history. Wars occur out of chains of often highly improbable events, and any theory of war prevention through crisis stability, in order to be responsible, has to provide a hedge against the possibility of a crisis erupting into war (a Soviet leadership may not be deterred when U.S. crisis stability theorists say it should). As is developed below, the case against MX/MPS on crisis (in)stability grounds is wrong, and logically can be demonstrated to be wrong.

It is wrong in two fundamental aspects: first, as to what should constitute an inherently stabilizing strategic posture in an acute crisis, and second, as to its (lack of) total integrity on the conflict continuum. There should be no need to face this problem in practice, but in principle, at least, one should be willing to face the possibility that there might be some tension between the strategic desiderata for a crisis-stabilizing posture and the strategic desiderata for the post-crisis prosecution of a war—and it is not a self-evident truth that considerations of the former should totally dominate the latter (if only because, although the prevention of war is preferable to the prosecution of war, there can be no guarantee that war might not occur anyway).[4] A strategic nuclear posture and its guiding doctrine should be designed for time of war as well as time of peace.

Preeminent among the reasons why crisis stability criteria continue to pose potentially deadly threats to some strategic weapon (and weapon-performance improvement) programs is the still very popular belief that a central war can only be lost by both sides. If that belief has real merit, then it is at least not unreasonable to devote the lion's share of one's care and attention to war prevention. Almost certainly it is the case that neither superpower would be likely to emerge from a central war and attain the object of war as defined by Basil Liddell Hart: "a better peace."[5] The prewar situation would be preferable to the postwar.

But this is not to claim that an outcome describable as victory would be impossible. Great Britain did not obtain "a better peace," in the Liddell Hart sense, in 1918 and in 1945, but the somewhat hollow attainment of victory was still preferable to, and clearly distinguishable from, defeat. Some very popular beliefs notwithstanding, it is less than obvious that the kind of damage that probably would be suffered in a central nuclear war need preclude the validity of the traditional distinction between winning and losing. Even the CIA, which has no known vested interest in minimizing estimates of the damage likely to be suffered by the U.S.S.R., has predicted that under favorable conditions (which would not totally be a matter of luck), Soviet casualties could be kept down to "the low tens of millions," of which only half would be fatalities.[6] This is a horrific price tag, but it is probably lower than that suffered in World War II and should constitute well under 10 percent of the total population.

It is sensible to be skeptical concerning low or very low estimates of damage in a central war, but it is no less sensible to remember that the deterrent effect that really matters when Western theorists address the issue of crisis stability is the deterrent effect in Soviet minds. As a continental power, the Soviet Union knows that war against a first-class adversary invariably is a very expensive undertaking. Aside from the point that war could occur for reasons far removed from rational strategic calculation, it is not enough to repose one's hopes for war prevention upon the preservation of a capability to inflict catastrophic damage on the Soviet Union. Soviet leaders almost certainly expect to suffer catastrophic damage in war (they may hope for the best, but they prepare for much of the worst); the real issue is whether they expect to suffer intolerable damage.[7] Catastrophe could come in different orders of magnitude.

The argument against MX/MPS on crisis stability grounds has considerable difficulty coping with the clear evidence provided by Soviet programs in the 1970s. Can a theory of stability be valid if the opponent gives every appearance of not sharing it?[8] Stability, after all, is supposed to inhere in a relationship. Various explanations have been offered for the dynamics of the Soviet weapon acquisition process, but there is a very widespread consensus in the West on one critical fact: Soviet weapon development and procurement clearly has not been

inhibited by crisis stability considerations. Most participants in the
U.S. defense debate agree that the Soviet Union already poses an
unacceptably large threat to the prelaunch survivability of U.S.
ICBM's in silos. Also, it is generally agreed that Soviet strategic
doctrine calls for the physical protection or evacuation of Soviet
citizens from areas of high nuclear risk. Although there is debate in
the United States over Soviet prowess, the Soviet ambition to assure
the survival of its domestic assets is beyond dispute. To be direct,
the Soviets do not share our still dominant theory of strategic
stability. They do not believe it is stabilizing for (Soviet) people to
be vulnerable and for (Soviet) strategic weapons to be invulnerable. [9]

The fundamental crisis stability case against MX/MPS has been
expressed in a State Department publication:

> It is also U.S. policy to maintain strategic forces of
> sufficient strength, diversity, and survivability that the
> Soviets will not have an incentive to strike first in a
> crisis. Consistent with this view <u>it is our policy not to
> deploy forces which so threaten the Soviet retaliatory
> capability that they would have an incentive to strike first
> to avoid losing their deterrent force.</u> [10] (Emphasis added.)

The document proceeds with the curious claim and qualification
that "this policy is contingent on similar Soviet restraint." This logic
is curious, to say the least. The United States will not deploy forces
that threaten the Soviet retaliatory capability (because the State Department believes that such a capability provides an incentive to preempt),
but should the Soviets go that route, the United States might follow
them—thereby (with this logic) providing the Soviets with just such an
incentive to strike first. The moral of this story probably is that
strategic theory should be left to the professionals and not entrusted
to the State Department. [11]

The crisis stability case against MX/MPS—that it would provide
the Soviet Union with a possibly irresistible incentive to strike first
in an acute crisis—collapses on all fronts when subjected to rigorous
investigation. First, MX deployment, even if of a base-line variety
(200 missiles 92 inches in diameter, weighing 192,000 pounds,
capable of an 8,000-pound payload that could accommodate 10 Mk12A
reentry vehicles), could pose a fatal first strike threat only to most
of the Soviet silo-housed ICBM force. This force does, admittedly,
contain the major share of Soviet strategic firepower, but the Soviet
retaliatory assets that such an MX deployment could not touch would
remain impressive. MX/MPS, however large the program, cannot
possibly threaten Soviet retaliatory capability (at least against
American urban areas).

What MX/MPS can and should threaten is the Soviet capability against American hardened military targets and, above all, the likelihood that political profit could be extracted from the threat in, or actual execution of, that capability. American arms controllers cannot have it both ways. They cannot be permitted to argue, simultaneously, that although the Soviet threat to Minuteman is indeed worrisome, such a threat is not synonymous with a threat to the U.S. strategic deterrent, and to argue that MX/MPS is an anathema because it must generate crisis instabilities through the threat it would pose to Soviet fixed-site ICBMs. Soviet throw weight is indeed heavily represented in the ICBM force, but the Soviet SLBM and bomber forces are not of trivial proportions.

Second, MX/MPS, if well designed in the details of its basing mode (which one must—and should—assume, given the length of time, degree of effort, and quality of technical talent devoted to it), is by definition innocent of a potential for generating crisis instabilities. MPS basing deters attack either through posing impossibly large fractionation demands upon the Soviet ICBM force or through requiring, for its (technical) defeat, so disproportionate an expenditure of Soviet missile payload (to that destroyed in the MPS) that only a Pyrrhic victory, at best, could ensue. By deterring attack upon itself, MX/MPS should deny the Soviets a rational first-strike option.

Even if MX poses a potentially fatal threat to Soviet ICBMs, against what U.S. target structure does a highly nervous Soviet leadership launch its missiles? MX deployment could be criticized if it were in a basing mode that the Soviets should be able to overwhelm at a reasonable cost—but the whole point of MPS basing is to deny this possibility. So where lies the crisis instability? If the Soviets cannot target MX/MPS with any hope of success, but are fearful that 200 (or more) 10-MIRVed MX ICBMs might essentially take out their ICBM force in silos, is it reasonable to argue that they would launch their ICBM force against American society in order to avoid its defeat on the ground by a U.S. first strike? This author would like to meet in debate anybody who would be so foolish as to advance such an argument. In this claim the Soviets would be opting for war, while leaving the U.S. ICBM force untouched in the first round, or beyond. This argument is simply nonsensical.

Third, the crisis instability case against MX/MPS is fragile on the ground that the system will not attain FOC overnight. If, for the sake of argument, it is assumed that the crisis instability case against MX/MPS is basically sound, how much lead time is the United States likely to provide the Soviet defense establishment to enable it to head off instability through the phasing out of silo-housed ICBM assets (and their deployment in a mobile mode on land, their

replacement by sea and airborne forces, or the conduct of SALT
negotiations to seek a solution via a cooperative route)?

Presuming no crash program for MX/MPS deployment, the
Soviet Union has been provided nine years' notice that the system is
coming: six years to IOC, and a further three years to FOC. If the
Soviets continue to deploy ICBMs in silos, even as an MX/MPS program approaches full maturity, then there has to be something very
deficient about Soviet strategic thinking (a possibility that, on the
evidence available, we should discount); there has to be something
fundamentally wrong with the U.S. theory of crisis stability; or
there must be something very wrong with the U.S. MPS deployment
scheme.

Fourth, considered in the context of the arguments above, MX/
MPS should be a major element in a U.S. theory of crisis and escalation control. For full escalation control, the United States needs both
to be able to protect its war-recovery assets and to have a grand
design for the defeat of the Soviet state. With MX/MPS deployment
the United States would be able to threaten to execute both small-
scale and even some selective attack options, secure in the knowledge
that the Soviets could not respond with a devastating hard-target
counterforce. Moreover, should MX/MPS pose a nearly total threat
to Soviet ICBMs in silos, while remaining untargetable itself, the
Soviet Union would have to be extremely chary of offering even a
close-to-matching reply to limited and selective U.S. nuclear use.

No one can guarantee the predictive validity of any postulated
escalation sequence, but with MX/MPS in place, an American president in dire straits and in search of strategic employment options that
are not too self-deterring in prospect, should be enormously grateful
for the escalation control (and initiative) possibilities that are yielded
by virtue of that deployment. However, as already suggested, MX/
MPS can be no panacea. In and of itself it makes possible much of
the design of a politically intelligent strategy, but it requires the cover
of active and passive defense of the homeland for the United States to
have a deterrent with real integrity. [12]

11

DOMESTIC CHALLENGES

IS MX A GROSS EXAMPLE OF A PUBLIC WORKS PROJECT OF BENEFIT TO THE MILITARY-INDUSTRIAL COMPLEX?

The associate contractors undoubtedly will profit from the MX/MPS program.[1] However, there is nothing in the MX/MPS program of which the contributing companies should be ashamed. MX/MPS is desperately needed on political and military grounds— there is simply no case for arguing that it is a boondoggle for the defense industry. The plain fact that MX/MPS is good financial news for the U.S. aerospace and construction industries is neither here nor there. The United States and its allies and friends are in urgent need of the strategic capabilities of the MX missile system. A noticeable fraction of MX system costs derives not from the greed of defense contractors but, rather, from arms control sensitivities and from the fact that the program has already been expensively delayed by three years.

As this author has explained elsewhere, there is no such thing as the military-industrial complex (MIC).[2] The feast-and-famine cycles that beset U.S. defense industry are imposed, in toto, from without. The academic hypothesis of an American MIC cannot withstand exposure to the facts of the U.S. research, development, and procurement cycle.[3] While it cannot be denied that MX/MPS is a subject of major financial interest to aerospace and civil construction corporations, this author challenges any critic of MX/MPS to produce evidence suggesting that the program has been endorsed officially because of anticipated economic benefit. MIC-type inquiry directs attention to a nonsequitur in defense commentary. Although MX/MPS will be a source of profit for large defense contractors, the system has not been chosen with a view to maximizing corporate profits.

111

This author has closely followed the issues of the MX/MPS program for several years, but he has yet to hear of anything that would suggest an improper corporate-industrial voice in defense decision making. Indeed, forceful industry lobbying on behalf of MX has been close to nonexistent. Overall, it cannot be denied that the U.S. government has chosen to purchase MX/MPS on strategic-political grounds, and that defense contractors have played a close to insignificant role in that procurement decision.

When critics of MX write articles with titles like "MX: The Public Works Project of the 1980s,"[4] they imply, perhaps unintentionally, impropriety that has not been present. It is difficult to make sense of MIC charges. MIC theory can explain all aspects of defense policy (including all aspects of threat analysis), but, as with many conspiracy theories, it explains both everything and nothing. In essence, MIC theory holds that the armed forces (and probably the Defense Department as a whole) are in a malign, self-interested, functional alliance with the weapon manufacturers against the public interest. A similar coalition in the Soviet Union provides all of the external threat that the U.S. MIC needs in order to justify its products.[5]

The truth of the matter is that a weapon system like MX means jobs for tens of thousands of workers, profit for corporations both large and small, and requires close technical collaboration between manufacturers and the eventual operator—how could it be otherwise? President Carter's decision in 1979 to proceed with the MX missile in an MPS basing mode appeared to reflect a mix of domestic political, arms control bargaining, and assessed strategic need reasons. This author knows of no evidence suggesting that MX/MPS is intended primarily, or even as an important subsidiary motive, to profit particular interests. Profit is the engine of the American economic system: critics who do not like that fact should not confuse a system of which they do not approve with venality.

WON'T MX/MPS HAVE A VERY DISRUPTIVE EFFECT UPON THE AREA WHERE IT IS BASED?

It has been decided that the United States must, on strategic grounds (for deterrence), retain an ICBM force, and that that force must be deployed in a mobile—or at least moveable—configuration. There is an alternative, in that the president could choose actively to defend ICBMs in silos,[6] but for reasons that transcend the boundary of this discussion, the ABM treaty of 1972 continues to be held in high regard. However, by far the most efficient way to employ BMD in defense of ICBMs would be to marry active defense to a shell-game ICBM deployment. Such a scheme maximizes defensive leverage for a given investment.

The dilemma faced by the Air Force is that while the American people are adamant in insisting that their country not slip into a position of strategic inferiority, they tend to be less than enthusiastic over the prospect of having the essential military reply to the Soviet ICBM threat (MX/MPS) deployed close to them.[7] Most people want the benefit of MX/MPS (enhanced security), but there are few volunteers to suffer the local disruption. Also, there is the further dilemma that while it may be desirable on military and social grounds to deploy MX/MPS in areas of very low population density (relatively very few people are directly affected), such low population density means that the disruptive effect of the system is bound to be unusually severe.

The Air Force has done its job with admirable clarity. It has selected "the best" basing area according to six explicit criteria:

Physical security

Command, control, and communications

Reaction time

Missile accuracy

Land use

Feasibility of contiguous basing.

The Great Basin of Utah-Nevada is the clear "winner" on these criteria. Although the Air Force will do its best to minimize the burdens placed upon the local population, it cannot be denied that MX/MPS will bring much new economic activity, some challenging lifestyles (on the part of construction workers and the permanent security and operating force), and some problems of water usage. With goodwill, sensitivity, and careful planning, these problems can be minimized—and may even, in some cases (water, for a very important example), prove to be nonexistent. However, it is a political fact that local opposition to MX/MPS has grown markedly since 1979. In good part this opposition appears to reflect the judgment that it is unjust that Utah and Nevada should bear the entire burden of MX/MPS disruption.

Strategically, and as a matter of national defense decision making, there is no question that contiguous basing in the Great Basin is the sensible choice. The case against split basing comprises the following principal elements:

It is expensive, requiring duplication of facilities and inefficient use use of personnel and equipment.

It will require socially disruptive and time-consuming acquisition of private land (the land to be used in Utah-Nevada is all public).

It will probably delay the MX/MPS IOC date by a year. There is a legal requirement that a single environmental impact statement be filed for the entire MX/MPS system before a final decision on basing can be made.

The operational effectiveness of the entire system would be degraded by split basing. Split basing would have many of the MX missiles deployed in areas beyond unjammed medium-frequency radio communication range of other MXs. It is envisaged that after a Soviet attack, surviving elements of the MX force will communicate their status to each other—permitting rational retargeting.

Thus, split basing (say, between Utah-Nevada and the high plains of west-central Texas and eastern New Mexico) may be a domestic political necessity, but there is no dispute as to its strategic undesirability. [8]

The disruption that must be caused by the construction of 4,600 shelters, connecting roadways, and maintenance facilities will be minimized by such tactics as separating, insofar as possible, the transitory construction population from the local residents (through provision of trailer parks geographically well removed from local towns) and by completing the construction as rapidly as possible. Readers should be aware of the number of local residents likely to be affected very directly. The urban and rural population within the MX basing area is only 6,137. The smallness of this number minimizes the human scale of the disruption, but it also serves to maximize the impact on the individuals.

It is worth mentioning that although the operation of MX/MPS will require numbers of people on a scale noticeable in Utah-Nevada, local anxieties appear to be focused in particular upon the period of system construction. Local residents, in a socially very conservative region, are not enthusiastic about the prospective presence of tens of thousands of, by and large, single (or away from home) young construction workers.

There is no way in which a construction project on the scale of MX/MPS can be completed in a totally painless fashion. However, the scale and duration of the local social and economic problems can, and should, be affected positively by proper leadership. Above all else, perhaps, the local residents in Utah and Nevada—and elsewhere, if the Air Force is compelled to endorse split basing—should understand just how vital MX/MPS is to American security. From its very satisfactory experience with local residents in and around the Minuteman fields, the Air Force knows that it needs a friendly environment for deployment of a major weapon system.[9]

PART III
CONCLUSIONS

12

THE BALANCE OF PRUDENCE .

As must be abundantly clear from the discussion in Part II, this author believes that the various arguments that have been advanced in opposition to the MX missile system, and to the MPS basing mode, are not persuasive. However, it is altogether healthy that a weapon system as important as the MX ICBM should be subject to searching scrutiny. An interesting comparison may be drawn between the U.S. MX debate and the British Polaris follow-on debate. In the latter case a major weapon program, with enormous financial, political, and strategic implications, was decided upon by a government that, in the British tradition, viewed the decision to acquire the Trident SLBM essentially as lying within the realm of "executive discretion." Parliament and the country at large were informed of the decision, ex post facto.

The British Constitution and its standard operating procedures have many things to recommend them, but the prevailing ethos of secrecy and secretiveness on any subject that bears upon "national security" leads, inevitably, to a notably impoverished public debate. The poor quality of British defense policy in the twentieth century must, at least, in part, be attributed to the fact that British governments have not felt, or legally been, obliged to share many of their secrets with society at large. British government does not really want serious input to policy debates on defense issues from outside. So long as access to "the facts" is restricted to a handful of civil servants and military people, then, by definition, outside opinion must be ill-informed, and hence should not be taken seriously.

The American case of the MX ICBM offers almost a comical contrast with the British nondebate over Trident. [1] In the American defense community (an expansive concept embracing politicians, officials, think-tank analysts, journalists, defense contractor personnel, and academics), the standard operating procedures of leak and counterleak, and of early—and even premature— briefing of proposed weapon system details, ensure that most interests and schools of thought are very well armed with "selected facts" that suit their cases. In short, the American democratic process, with reference to major weapon system decisions, virtually guarantees the possibility of technological, strategic, or political filibuster.

There can be no question that the American way in defense policy formulation is infinitely preferable to the British. Ill-conceived, rash official proposals are rapidly exposed for the folly they are. Unlike the British case (as with Trident), the MX issue is being debated by an extraofficial American defense community of genuinely expert commentators. However, life offers few free goods—and this is true of the American way with defense program details. As noted earlier in this study, a point comes when it is prudent on strategic grounds to say "enough debate." The United States cannot deter Soviet leaders with weapon systems that are only on the drawing boards, subject to expert contention.

All of the important questions to be asked concerning the wisdom of procuring a major weapon system cannot be answered, definitively, in a timely fashion. Although this author believes that the United States should move very expeditiously to procure MX/MPS, he admits that questions, of varying degrees of importance, remain—and have to remain—unanswered in full. For example:

There is always the possibility that some group of weapon engineers will invent an ICBM (and basing) concept superior to MX/MPS.

The Soviet Union may surprise us unpleasantly with respect to the quality and quantity of its offensive and defensive arms pertinent to the prospective MX deployment era of 1986-2006.

It is just possible that the $30-35 billion (1980 dollars) to be invested in MX/MPS could produce better deterrent effect if applied to other defense programs.

This study has sought to expose the reader to the full detail of the more important aspects of the MX debate. In so doing, the risk has been accepted that more information has been provided than the reader will believe usable. However, a complex issue is a complex issue. The MX debate is not just about the modernization of the ICBM force, the environmental impact in Utah-Nevada, the implementation of particular (and controversial) war plans, the course of the strategic arms competition, the support of U.S. allies, or the future of the SALT process—to select only some of the major themes—it is about all of these elements, simultaneously.

This book treats the MX missile debate seriously as a strategic debate. Many people, of course, decline to view MX in strategic perspective. They see MX in terms of hospitals not built, virgin public lands appropriated and scarred, and further inflationary pressures. This book has had little to say to such people: their human concerns are applauded (and are shared by the author), but their sense of national priorities is deplored. In the best of all possible worlds the United States would not need MX ICBMs, or any missiles at all. However, "the Soviet threat," which has loomed large in these pages, is an unfortunate reality, and not a vision of unrepentant cold warriors.

The imminent vulnerability of the U.S. ICBM force (and of large fractions of other elements of the U.S. strategic force posture) could be the Achilles heel that brings down the entire edifice of Western defense. The MX ICBM, deployed survivably in multiple protective structures, far from being an arms race initiative by American "hawks," is a dangerously belated response to a clear and present danger. The United States needs to rebase its ICBM force because the Soviet Union has chosen to purchase a land-based missile force that places U.S. ICBMs at intolerable risk; in addition, the United States needs a highly accurate, large (by U.S. standards) ICBM if it is to pose the kind of threat to premium Soviet state assets that Soviet leaders are known to find maximally deterring.

Some of the journalistic hyperbole of recent months has encouraged a quite fallacious view of the arguments of the proponents of MX/MPS. To set the record straight, advocates of MX/MPS typically do not do the following:

Assert that MX/MPS is the perfect weapon for all time—instead, they argue that MX/MPS is necessary, and good enough, for the 1980s, 1990s, and some modest period thereafter

Assume that the U.S.S.R. will be so intimidated by the prospect of MX/MPS that it will reformulate its SALT negotiating strategy along agreeably cooperative lines

Assume that the U.S.S.R. will be incapable of procuring a serious threat to the system—rather, they argue that base-line MX/MPS (200 MX missiles in 4,600 shelters) has a growth potential for survivability, including BMD options, that will defeat any Soviet bid to overwhelm the system

Assume that any nuclear war could be limited—but they do argue that the prospects for war limitation (and for the deterrence of war) will be enhanced by U.S. procurement of MX/MPS.

Critics of MX/MPS have endeavored, with some success, to convey the impression that MX is an overly complicated behemoth—a

Rube Goldberg machine. This view does not withstand exposure to the facts. Although the debate over MX/MPS is complex in the factors that it has embraced, the military system in question is far from complex in its concept of operation. A linear grid of 23 shelters will contain a single MX missile, to be shuffled occasionally, deceptively, from one shelter to another. Also, possibly, this shuffle might be effected on a basis of urgency in a period of tension. This is perhaps complex by comparison with a stationary ICBM suspended in a silo, as at present, but it is very simple when compared with the readiness moves of other elements of the armed forces.

The virtue of debate becomes a vice if it delays decision beyond the point where a strategic response is required. It is now time for action on MX/MPS, not for further deliberation.

NOTES

INTRODUCTION

1. For an exposition of the strategy questions pertinent to MX, see Colin S. Gray, Strategy and the MX, Critical Issues series (Washington, D.C.: Heritage Foundation, 1980).

2. For example, though many people will disagree, this author believes that neither the Safeguard ABM system (in 1969-70) nor the B-1 (in 1977) was assessed properly on its strategic merits, either by the U.S. government or by many outside critics. This is not to suggest that either system did not merit technical criticism; it is to claim that the ABM, in 1969-70, was a symbolic issue for both generically pro- and anti-defense groups, while in 1977 the B-1 was virtually the symbolic issue with respect to which President Carter was compelled to prove his good faith to his liberal constituency.

3. See Adam Clymer, "Behind Every Defense Policy There Lurks a Political Idea," New York Times, August 24, 1980, p. E4.

4. See William H. Kincade, "A Farewell to Arms Control?" Arms Control Today 10, no. 1 (January 1980): 1-5.

CHAPTER 1

1. The "great ABM debate" may be approached most economically through Abram Chayes and Jerome B. Wiesner, eds., ABM: An Evaluation of the Decision to Deploy an Antiballistic Missile System (New York: Harper and Row, 1969); and Johan J. Holst and William Schneider, Jr., Why ABM? Policy Issues in the Missile Defense Controversy (New York: Pergamon, 1969). An excellent retrospective analysis of the ABM debate is Keith B. Payne, The BMD Debate: Ten Years After, HI-3040/2-P (Croton-on-Hudson, N.Y.: Hudson Institute, 1980).

2. See Francis Hoeber, Slow to Take Offense: Bombers, Cruise Missiles, and Prudent Deterrence (Washington, D.C.: Center for Strategic and International Studies, Georgetown University, 1977).

3. The manned bomber issue was injected very directly into the 1980 election campaign when the Carter administration decided to leak details of a new "stealth" aircraft—a "black" weapon program designed to produce a vehicle very substantially immune to radar detection. See Walter S. Mossberg, "'Invisible' Plane Disclosure Looks

123

Certain to Affect Soviet Relations, U.S. Politics," Wall Street Journal, August 25, 1980, p. 6. Beyond the obvious element of election-year politics, the disclosure of "stealth" technology and the major claims advanced for it by Secretary of Defense Harold Brown seem likely to have been motivated by Mr. Carter's strong desire to head off congressional sentiment in favor of building either the B-1 or a close B-1 derivative. See Clarence A. Robinson, Jr., "Multipurpose Bomber Advances," Aviation Week and Space Technology 113, no. 5 (August 4, 1980): 16-18. Unsurprisingly, and possibly correctly, the administration was accused both of overselling the new technology and of risking the compromising of its effectiveness through such early disclosure (thereby permitting the U.S.S.R. a very long leadtime to seek countermeasures).

4. Department of Defense Annual Report, Fiscal Year 1981 (Washington, D.C.: U.S. Government Printing Office, 1980), p. 14.

5. Useful commentaries include Jacek Kugler and A. F. K. Organski, with Daniel Fox, "Deterrence and the Arms Race: The Impotence of Power," International Security 4, no. 4 (Spring 1980): 105-38; and Kendall D. Moll and Gregory M. Luebbert, "Arms Race and Military Expenditure Models: A Review," Journal of Conflict Resolution 24, no. 1 (March 1980): 153-85.

6. A classic statement of the simple "action-reaction" theory is George W. Rathjens, "The Dynamics of the Arms Race," Scientific American 220, no. 4 (April 1969): 15-25. For detailed critical reviews of such theories, see Albert J. Wohlstetter, Legends of the Arms Race (Washington, D.C.: U.S. Strategic Institute, 1975); and Colin S. Gray, The Soviet-American Arms Race (Farnborough, England: Saxon House/D.C. Heath, 1976).

7. Tentative, though unmistakable, official acceptance of this proposition may be found in Harold Brown, Department of Defense Annual Report, Fiscal Year 1981, pp. 2-3; and his speech at Naval War College, Newport, R.I., August 20, 1980.

8. See Kincade's article "Will MX Backfire?" Foreign Policy no. 37 (Winter 1979-80): 43-58.

9. On assuming office in January 1977, the Carter administration generally reflected the pessimistic conclusions of the latest National Intelligence Estimate (NIE) on the Soviet Union that it inherited ("National Estimate of Soviet Strategic Capabilities and Objectives, NIE 11/8," signed by the director of the Central Intelligence Agency [George Bush] on January 7, 1977). This NIE claimed that the U.S.S.R. was determined to achieve strategic superiority and to be in a position to win a nuclear war with the United States by 1985. But the 1976 NIE, conclusions aside, offered projections for Soviet strategic forces that subsequently (in 1978—when the next NIE on Soviet strategic forces was prepared, and again in 1979) were seen to have been far

too optimistic. To be fair to Mr. Carter, his 1977 decisions not to proceed with B-1 production, and to delay engineering development of MX, might in part be traced to the erroneous NIE that he inherited. See "Strategic Cuts Laid to Faulty Intelligence," Aviation Week and Space Technology 112, no. 8 (February 25, 1980): 19-20.

10. Dr. Kissinger asserts in his memoirs that he viewed SALT I, inter alia, as an opportunity for the United States to "catch up": "We needed the agreement if we wanted to catch up in offensive weapons." White House Years (Boston: Little, Brown, 1979), p. 1245. Kissinger's version may be correct, but readers would do well to attend carefully to Admiral Zumwalt's accounts of Kissinger's (privately expressed) deep pessimism—they have the ring of truth. Elmo Zumwalt, On Watch (New York: Quadrangle, 1976), esp. p. 319.

11. Bernard Brodie, "On the Objectives of Arms Control," International Security 1, no. 1 (Summer 1976): 17-36.

12. Brown, Department of Defense Annual Report, Fiscal Year 1981, p. 41.

13. Henry Kissinger, press conference in Vladivostok, November 24, 1974.

14. "Essential equivalence" is a slippery concept to define—readers are recommended to sample the year-by-year semantic variations in Defense Department reports.

15. The phenomenon has been described brilliantly and mercilessly in Kenneth L. Adelman, "Rafshooning the Armageddon: Selling SALT," Policy Review no. 9 (Summer 1979): 85-102.

16. See N. H. Gibbs, Grand Strategy: vol. 1, Rearmament Policy (London: Her Majesty's Stationery Office, 1976), pp. 3-6, 55-87.

17. The most effective hard-target killer in the American arsenal, the Mk12A warhead, installed on 300 Minuteman III ICBMs, can achieve a single-shot kill probability against (most) Soviet ICBM silos only on the order of 0.29. Two (warheads)-on-one (silo) targeting is infeasible because of the shortage of Mk12A warheads.

18. A fact that no responsible person disputes. NATO's program for modernizing its long-range theater nuclear forces with 572 ground-launched cruise missiles and Pershing IIs is belated policy recognition of this fact.

19. "The Illusions of Distance," Foreign Affairs 46, no. 2 (January 1968): 242-55.

20. See William Schneider, Jr., "Soviet Military Airlift: Key to Rapid Power Projection," Air Force Magazine 63, no. 3 (March 1980): 80-86.

21. The Soviet Union has yet to begin actually laying down a nuclear-powered aircraft carrier, but the balance of intelligence indicators suggests very strongly that such a development is imminent.

22. The Soviet airlift to Ethiopia in January 1978 was a particularly impressive demonstration of the increasingly muscular trend in Soviet intervention policy. On the possible significance of the Soviet operation to rescue the Ethiopian government, see Edward N. Luttwak, "After Afghanistan, What?" Commentary 69, no. 4 (April 1980): 40-49.

23. John Erickson, "The Soviet Military System: Doctrine, Technology and 'Style,' " in John Erickson and E. J. Feuchtwanger, eds., Soviet Military Power and Performance (Hamden, Conn.: Archon, 1979), p. 25.

24. Recent U.S. NIEs show a vast difference between a SALT-limited Soviet strategic warhead inventory and an inventory developed free of SALT limitations. For a rare public exposure of some alleged NIE figures, see Richard Burt, "Soviet Nuclear Edge in Mid-80's Is Envisioned by U.S. Intelligence," New York Times, May 13, 1980, p. A12.

CHAPTER 2

1. As this author has developed in a separate study, each society is attracted to strategic and political theories that reflect its world view and its interests. As fundamentally "satisfied," status quo powers, NATO members naturally are attracted to the concepts of order and stability. For an interim statement of this thesis, see Colin S. Gray, "Strategic Stability Reconsidered," Daedalus 109, no. 4 (Fall 1980): 135-54.

2. The Soviet Union probably devotes very close to 15 percent of its GNP to defense functions—quite narrowly interpreted. See William T. Lee, Soviet Defense Expenditures in an Era of SALT (Washington, D.C.: U.S. Strategic Institute, 1979).

3. See the trenchant, brief analysis in William E. Griffith, "Super-Power Relations After Afghanistan," Survival 22, no. 4 (July/August 1980): 146-51. For a comprehensive survey see Herbert Goldhammer, The Soviet Soldier: Soviet Military Management at the Troop Level (New York: Crane, Russak, 1975).

4. Department of Defense Annual Report, Fiscal Year 1980 (Washington, D.C.: U.S. Government Printing Office, 1979), p. 6.

5. The "quick fix" route is argued in detail in William R. Van Cleave and W. Scott Thompson, eds., Strategic Options for the Early Eighties: What Can Be Done? (New York: National Strategy Information Center, 1979). Also see William R. Van Cleave, "Quick Fixes to U.S. Strategic Nuclear Forces," in W. Scott Thompson, ed., National Security in the 1980's: From Weakness to Strength (San Francisco: Institute for Contemporary Studies, 1980), ch. 5.

6. Henry Kissinger, "The Future of NATO," Washington Quarterly 2, no. 4 (Autumn 1979): 4.

7. U.S. Senate, Committee on Foreign Relations, The SALT II Treaty, Hearings, Pt. 3, 96th Cong., 1st sess. (Washington, D.C.: U.S. Government Printing Office, 1979), pp. 164-65.

8. However, we do know that Soviet analysts tend to study "the correlation of forces" rather than the more narrow (Western) military or strategic balance. See Benjamin S. Lambeth, "The Political Potential of Soviet Equivalence," International Security 4, no. 2 (Fall 1979): 22-39; and Vernon V. Aspaturian, "Soviet Global Power and the Correlation of Forces," Problems of Communism 29 (May-June 1980): 1-18.

9. Leslie H. Gelb, "A Glass Half Full," Foreign Policy no. 36 (Fall 1979): 22.

10. See Norman Stone, The Eastern Front, 1914-1917 (London: Hodder and Stoughton, 1975), ch. 1.

11. This is not to deny that, as early as November 5, 1937, Hitler told his military chiefs that Germany's problem of lebensraum would have to be solved by 1943-45, at the latest, because "at that time the law of obsolescence would begin to blunt the efficiency of Germany's armed forces. . ." Gordon A. Craig, Germany, 1866-1945 (New York: Oxford University Press, 1978), p. 698. Soviet policy, unlike German policy between 1933 and 1945, has never been conducted as a high-risk adventure.

12. To be deployed, in their air-launched version, over the period 1982-90 (on the relaxed Carter schedule).

13. For contrasting views on this subject, see Richard Pipes, "Why the Soviet Union Thinks It Could Fight and Win a Nuclear War," Commentary 64, no. 1 (July 1977): 21-34; and Aleksandr Solzhenitsyn, "Misconceptions about Russia Are a Threat to America," Foreign Affairs 58, no. 4 (Spring 1980): 797-834.

14. Leslie Gelb and Richard Ullman, "Keeping Cool at the Khyber Pass," Foreign Policy no. 34 (Spring 1979): 7.

15. Secretary of Defense Harold Brown acknowledged the problem in his speech at the Naval War College, Newport, R.I., on August 20, 1980: "In the future, Soviet military programs could, at least potentially, threaten the survivability of each component of our strategic forces. For our ICBM's that potential has been realized, or close to it." (Emphasis added.)

16. Stephen Rosen, "Safeguarding Deterrence," Foreign Policy no. 35 (Summer 1979): 123.

17. Robert Jervis, the commentator in question, published the paper under the title "Why Nuclear Superiority Doesn't Matter," Political Science Quarterly 94, no. 4 (Winter 1979/80): 617-33.

18. See Colin S. Gray, "Nuclear Strategy: The Case for a Theory of Victory," International Security 4, no. 1 (Summer 1979): 54-87.

19. On the fate of "superiority" in U.S. defense thinking in the 1960s and early 1970s, see Jerome H. Kahan, Security in the Nuclear Age: Developing U.S. Strategic Arms Policy (Washington, D.C.: Brookings Institution, 1975), pp. 106-09, 144-49.

20. See Colin S. Gray, "The Strategic Forces Triad: End of the Road?" Foreign Affairs 56, no. 4 (July 1978): esp. 772-75.

21. Kissinger, "The Future of NATO," p. 6.

22. See Colin S. Gray, Strategy and the MX, Critical Issues series (Washington, D.C.: Heritage Foundation, 1980), pp. 51-53.

23. Presidential Directive (PD) 59 on U.S. targeting strategy—publicly revealed in Richard Burt, "Carter Said to Back a Plan for Limiting Any Nuclear War," New York Times, August 6, 1980, pp. A1, A6—with its reported emphasis upon countermilitary and counter-political control targeting, cannot be effected without MX.

24. In a radio talk John Erickson said: "...the Soviet [military] command was not in business to amplify Soviet vulnerabilities and turned instead to developing a military/battlefield view of the nuclear scene as opposed to the managerial models elaborated in the United States, the sort of semantic nonsense which hid war behind a phrase like violent bargaining." Transcript of talk, entitled "The Soviet View of Nuclear War," broadcast on BBC Radio 3, June 19, 1980, p. 5. A classic exposition of what Erickson referred to as "semantic nonsense" is Thomas C. Schelling, Arms and Influence (New Haven: Yale University Press, 1966), esp. ch. 1.

25. McGeorge Bundy, "To Cap the Volcano," Foreign Affairs 48, no. 1 (October 1969): 13.

26. See Morton H. Halperin, "Clever Briefers, Myopic Analysts, and Crazy Leaders," unpublished paper prepared for the Aspen Arms Control Summer Study, June 1973.

27. Lion V. Sigal, "Rethinking the Unthinkable," Foreign Policy no. 34 (Spring 1979): 39.

28. Prepared statement in U.S. Senate, Committee on Foreign Relations, The SALT II Treaty, Pt. 3, p. 169, fn. 4.

CHAPTER 3

1. For example, see Bruce A. Smith, "Opposition to MX Seen Growing in Utah," Aviation Week and Space Technology 112, no. 22 (June 2, 1980): 20-21.

2. See Richard L. Garwin, "Launch Under Attack to Redress Minuteman Vulnerability?" International Security 4, no. 3 (Winter 1979/80): 117-39.

3. For example, this author actively advocated expeditious development of a deceptively based MX ICBM as early as 1974—an

advocacy that eventually saw the light of day in The Future of Land-Based Missile Forces, Adelphi Papers no. 140 (London: International Institute for Strategic Studies, 1977).

4. William R. Graham and Paul H. Nitze, "Viable U.S. Strategic Missile Forces for the Early 1980's," in W. R. Van Cleave and Scott Thompson, eds., Strategic Options for the Early Eighties (New York: National Strategy Information Center, 1979), ch. 6; also see prepared statement of William R. Van Cleave, "Solutions to ICBM Vulnerability," before U.S. Senate, Committee on Appropriations, Subcommittee on Military Construction, May 7, 1980.

5. The issue of "quick fixing" the ICBM vulnerability problem by means of Minuteman rebasing in an ALPS-MAPS configuration is a matter in urgent need of policy attention. This author has examined the detailed technical case pro and con Minuteman rebasing, and remains unconvinced that the United States should pursue this option. However, that is largely a technical judgment. There is no dispute between the rebasing advocates and opponents concerning the urgency of the need to solve the ICBM vulnerability problem.

6. Bruce A. Smith, "USAF Changes MX Missile Launch Mode," Aviation Week and Space Technology 112, no. 11 (March 17, 1980): 20-21; "MX Basing Decisions Clear Way for Design Advances," Aviation Week and Space Technology 112, no. 24 (June 16, 1980): 132-34; and "Alternative MX Basing Poses Problems," Aviation Week and Space Technology 112, no. 26 (June 30, 1980): 22-23.

7. See Edward N. Luttwak, "SALT and the Meaning of Strategy," Washington Review 1, no. 2 (April 1978): 16-28; and "The American Style of Warfare and the Military Balance," Survival 21, no. 2 (March/April 1979): 57-60.

8. See Harold Brown, Department of Defense Annual Report, Fiscal Year 1980 (Washington, D.C.: U.S. Government Printing Office, 1979), pp. 77-78.

9. This anxiety is no longer valid, given the counterforce and countercontrol theses advanced by Harold Brown in his explanation of the "countervailing strategy" endorsed in P.D. 59. See Brown, speech at the Naval War College, Newport, R.I., August 20, 1980.

10. Many people dislike being reminded that the Department of Defense is in the business of war planning.

11. See Fritz Ermath, "Contrasts in American and Soviet Strategic Thought," International Security 3, no. 2 (Fall 1978): 138-55.

12. See Colin S. Gray, "Strategic Stability Reconsidered," Daedalus 109, no. 4 (Fall 1980): 135-54.

13. A good example of such recognition is Robert Jervis, "Why Nuclear Superiority Doesn't Matter," Political Science Quarterly 94, no. 4 (Winter 1979/80): 630.

14. A point well made in Henry S. Rowen: "Formulating Strategic Doctrine," in Commission on the Organization for the Conduct of Foreign Policy, vol. 4, app. K, Adequacy of Current Organization: Defense and Arms Control (Washington, D.C.: U.S. Government Printing Office, 1975), pp. 219-34; and "The Evolution of Strategic Nuclear Doctrine," in Laurence Martin, ed., Strategic Thought in the Nuclear Age (Baltimore: Johns Hopkins University Press, 1979), pp. 131-56.

15. Common sense can be in short supply on this subject. Some commentators insist that the Soviet Union will misread U.S. strategic selectivity, in the context of a new emphasis on counterforce targeting, as a determination to achieve a war-winning capability (as if that were a crime in Soviet eyes). See Tom Wicker, "Terror in Disguise," New York Times, August 24, 1980, p. E21.

16. However, because of the different ICBM deployment practices of the United States and the U.S.S.R., it cannot be denied that it would be very difficult for a United States determined to minimize unwanted collateral damage to strike effectively at the Soviet ICBM force. This point is very well made in Desmond Ball, "Research Note: Soviet ICBM Deployment," Survival 22, no. 4 (July/August 1980): 167-70. This issue is really moot because the United States, through most of the 1980s, will lack the ability to pose a very substantial threat to Soviet ICBM silos.

17. On only one occasion has a professional group sought to pass judgment on the quality of public policy analysis—over rival presentations on the merits of the Safeguard ABM system. See Operaions Research Society of America, "The Nature of Operations Research and the Treatment of Operations Research Questions in the 1969 Safeguard Debate," Operations Research 19, no. 5 (September 1971): 1123-58; and Operations Research 20, no. 1 (January-February 1972): 205-45. This experiment was not a great success. Even though the investigation was limited to what could reasonably be identified as operations research aspects of the ABM debate, it was argued—with some reason—that major defense questions far transcend issues accessible to operations research skills.

18. See the comprehensive survey of this technology in Aviation Week and Space Technology 113, nos. 4 and 5 (July 28 and August 4, 1980).

19. Robert Kaiser and Walter Pincus, "The Doomsday Debate: 'Shall We Attack America?'" Parameters 9, no. 4 (December 1969): 79-85. If Kaiser and Pincus have some sense of history and are open to a little healthy mockery, they may appreciate that their "briefing" could have been set in Tokyo in November 1941 (not to mention Berlin in July 1914)—a thought they should find disturbing.

20. As, for example, in Stanley Sienkiewicz, "Observations on the Impact of Uncertainty in Strategic Analysis," World Politics 32, no. 1 (October 1979): 90–110.

21. Bearing directly on this subject, one of the most important statements of the nuclear age was made by Fred Ikle in "The Prevention of Nuclear War in a World of Uncertainty," speech given at the Joint Harvard/MIT Arms Control Seminar, Cambridge, Mass., February 20, 1974. (Mimeographed.)

22. As, for example, in William Kincade, "Will MX Backfire?" Foreign Policy no. 37 (Winter 1979–80): 47–51.

23. Periodic, publicized failures of NORAD's strategic early warning system— as occurred several times in late 1979 and 1980— are the kinds of events that fuel more generalized charges of technical incompetence.

CHAPTER 4

1. Harold Brown, speech at the Naval War College, Newport, R.I., August 20, 1980; Henry Kissinger, "The Future of NATO," Washington Quarterly 2, no. 4 (Autumn 1979): 3–17; and McGeorge Bundy, "Maintaining Stable Deterrence," International Security 3, no. 3 (Winter 1978/79): 11–13.

2. Harold Brown certainly displayed no great anxiety about Minuteman vulnerability when he assumed office in 1977.

3. Fratricide is the phenomenon whereby nuclear explosions create such turbulent local conditions that other incoming warheads are damaged, destroyed, or are made to deviate from their intended trajectories. The early to mid-1970s' belief that Minuteman would be protected by the fratricide phenomenon was eroded by the facts of unexpectedly rapid Soviet improvements in the accuracy of their fourth-generation SS-18 and SS-19 ICBMs.

4. See the testimony of James Schlesinger in U.S. Senate, Committee on Foreign Relations, Subcommittee on Arms Control, International Law, and Organization, Briefing on Counterforce Attacks, Hearings, 93rd Cong., 2nd sess. (Washington, D.C.: U.S. Government Printing Office, 1974); Paul Nitze, "Deterring Our Deterrent," Foreign Policy 25 (Winter 1976/77): 195–210; and "Assuring Strategic Stability in an Era of Detente," Foreign Affairs 54, no. 2 (January 1976): 207–32.

5. McGeorge Bundy, "To Cap the Volcano," Foreign Affairs 48, no. 1 (October 1969): 10; and "The Future of Strategic Deterrence," Survival 21, no. 6 (November/December 1979): 268–72.

6. Bundy, "Maintaining Stable Deterrence," p. 15.

7. Bundy, "The Future of Stable Deterrence," p. 272.

8. For example, see Wayne Biddle, "The Silo Busters: Misguided Missiles: The MX Project," Harper's 259, no. 1555 (December 1979): 45. Also, see Colin S. Gray, The Future of Land-Based Missile Forces, Adelphi Papers no. 140 (London: International Institute for Strategic Studies, 1977). Obvious as it may be, it is worth remembering that although the strategic forces "triad" was substantially the unplanned outcome of successful, competing weapon programs, it does not follow, ipso facto, that that outcome was necessarily unfortunate. Governments can do some foolish things deliberately, and some wise ones without much careful prior consideration.

9. See George H. Quester, Offense and Defense in the International System (New York: John Wiley, 1977).

10. Assuming a 20-year active operational lifespan for an MX ICBM that achieves its IOC in 1986.

11. Colin S. Gray, "Strategic Stability Reconsidered," Daedalus 109, no. 4 (Fall 1980): 135-54.

12. Testimony of James Schlesinger in U.S. Senate, Committee on Foreign Relations, Subcommittee on Arms Control, International Law, and Organization, Briefing on Counterforce Attacks, p. 3; and Kissinger, "The Future of NATO," esp. pp. 6-7.

13. In the words of John Erickson, "The Soviet strategic missile forces are organized into armies, brigades and regiments, geared to salvo firing— in short, it is a battlefield deployment of strategic weapons, 'nuclear guns', if you like, aimed at the enemy in order to fight a 'counter nuclear battle'...." "The Soviet View of Nuclear War," broadcast on BBC Radio 3, June 19, 1980, p. 8.

14. Harold Brown, Department of Defense Annual Report, Fiscal Year 1981 (Washington, D.C.: U.S. Government Printing Office, 1980), pp. 65-68.

15. See Arthur J. Alexander, Decision-making in Soviet Weapons Procurement, Adelphi Papers nos. 147-48 (London: International Institute for Strategic Studies, Winter 1978/79); Karl F. Spielmann, Analyzing Soviet Strategic Arms Decisions (Boulder, Colo.: Westview, 1978); and Norman Friedman, "The Soviet Mobilization Base," Air Force Magazine 62, no. 3 (March 1979): 65-71.

CHAPTER 5

1. France is very seriously considering deploying a new SX mobile missile on tractor-trailers on highways. Edward W. Bassett, "France to Modernize Nuclear Forces," Aviation Week and Space Technology 112, no. 24 (June 16, 1980): 269.

2. See Colin S. Gray, The Future of Land-Based Missile Forces, Adelphi Papers no. 140 (London: International Institute for Strategic Studies, 1977).

3. With synergistic BMD deployment— if needed.

4. See Paul K. Davis, Land-Mobile ICBM's: Verification and Breakout, ACIS Working Paper no. 18 (Los Angeles: Center for International and Strategic Affairs, UCLA, 1980).

5. See Colin S. Gray, Ballistic Missile Defense: A New Debate for a New Decade, HI-3160/3-P (Croton-on-Hudson, N.Y.: Hudson Institute, 1980).

6. See Donald Brennan's comment in U.S. Senate, Committee on Foreign Relations, The SALT II Treaty, Hearings, Pt. 4, 96th Cong., 1st sess. (Washington, D.C.: U.S. Government Printing Office, 1979), p. 394.

7. See Malcolm Wallop, "Opportunities and Imperatives of Ballistic Missile Defense," Strategic Review 6, no. 4 (Fall 1979): 13-21.

8. This author holds an agnostic position on the promise of weaponized lasers. At the present time the U.S. defense community is quite sharply divided into "true believers" and no less dedicated skeptics. The best technical advice on the subject appears to conclude that the strategic promise of laser BAMBI is not high, but experts have been known to be wrong.

9. Assuming, reasonably, that the Mk12A RV will be the one selected for MX.

10. Some advocates of "quick fixing" the silo vulnerability problem are urging just such a development. This author, while approving of their motives, is profoundly skeptical of their timetable (to be completed by late calendar year 1984).

11. This is not a very popular option, but it certainly is technically feasible— albeit at the price of a very considerable payload penalty to be traded for the required range.

12. As Harold Brown implied in his speech at the Naval War College, Newport, R.I., August 20, 1980. Also see Desmond Ball, Deja Vu: The Return to Counterforce in the Nixon Administration (Los Angeles: California Seminar on Arms Control and Foreign Policy, 1974); and Developments in U.S. Strategic Nuclear Policy Under the Carter Administration, ACIS Working Paper no. 21 (Los Angeles: Center for International and Strategic Affairs, UCLA, 1980).

13. Testimony in U.S. Senate, Committee on Foreign Relations, Subcommittee on International Organization and Disarmament Affairs, Strategic and Foreign Policy Implications of ABM Systems, Hearings, Pt. 3, 91st Cong., 1st sess. (Washington, D.C.: U.S. Government Printing Office, 1969), p. 672.

14. The shift against a counterforce emphasis in declared targeting doctrine, after the mid-1960s, was very much a coerced one. Many of the documents supporting NSDM 242 (of April 1974), the key targeting directive of the Schlesinger era, make it very clear that counterforce targeting (of all kinds) had, for many years, been relatively unattractive even on strictly military grounds. On the Schlesinger shift see Lynn Etheridge Davis, Limited Nuclear Options: Deterrence and the New American Doctrine, Adelphi Papers no. 121 (London: International Institute for Strategic Studies, 1975/ 76). On the evolution of the strategic nuclear balance, see John M. Collins, U.S.-Soviet Military Balance: Concepts and Capabilities, 1960-1980 (New York: McGraw-Hill, 1980), pt. III.

15. A nearly definitive commentary remains Johan J. Holst, "Strategic Arms Control and Stability: A Retrospective Look," in J. J. Holst and William Schneider, Jr., eds., Why ABM? (New York: Pergamon, 1969), ch. 12.

16. Colin S. Gray, "Strategic Stability Reconsidered," Daedalus 109, no. 4 (Fall 1980): 135-54.

17. See Thomas Schelling's exploration of this idea in his The Strategy of Conflict (Cambridge, Mass.: Harvard University Press, 1960), ch. 9.

18. Some Western commentators choose to take comfort in the thought that Soviet operational nuclear strategy may differ markedly from Soviet military doctrine. For example, see Elizabeth Pond, "Deterring Nuclear War: 2: The Soviet View," Christian Science Monitor, August 27, 1980, pp. 12-14. This thought may be well based in fact, but there is no evidence to suggest that it is. "The basic [Western] hawk view" (ibid., p. 14) of Soviet strategy, as presented in Joseph D. Douglass, Jr., and Amoretta M. Hoeber, Soviet Strategy for Nuclear War (Stanford, Calif.: Hoover Institution Press, 1979), may be disturbing, but—based as it is on Soviet military texts— what grounds are there for believing that Douglass and Hoeber have not read Soviet operational intentions accurately?

19. See Colin S. Gray, Strategy and the MX, Critical Issues series (Washington, D.C.: Heritage Foundation, 1980), pp. 45-53.

20. A major reason why there are many technical doubts about BMD is that the United States has never operated BMD "in the field." The ABM treaty of 1972 does permit the United States to gain field operating experience at one ABM site— that license should have been exercised.

21. In late 1978-early 1979, the USAF strongly supported MX deployment in vertical shelters. A year later it was defending the president's preference for horizontal shelters. See (and contrast) General Lew Allen, Jr., chief of staff, USAF, letter to Melvin Price, chairman, Committee on Armed Services, U.S. House of Representa-

tives, December 29, 1978; and letter to John C. Stennis, chairman, Subcommittee on Defense, Committee on Appropriations, U.S. Senate, February 28, 1980.

22. "Okay, on with MX," The Economist, May 10, 1980, p. 12.

23. President Carter, "MX Missile System, Remarks on September 7th, 1979," Department of State Bulletin 79, no. 2032 (November 1979): 25-26.

24. See Alton K. Marsh, "Pentagon Selects MX Grid Plan," Aviation Week and Space Technology 112, no. 19 (May 12, 1980): 15-16.

25. Also, if the United States fires an MX missile, the Soviets— through MX closed clustering— would know, ipso facto, of 23 shelters that need not be targeted. See Desmond Ball, "The MX Basing Decision," Survival 22, no. 2 (March/April 1980): 64-65, fn. 27.

26. 600 psi resistance is the minimum design objective for the horizontal shelters— actual psi resistance may be much higher.

27. See Richard Burt: "Likelihood of SALT's Demise Changes the Strategic Options," New York Times, March 23, 1980, p. E4; "MX in Search of a Home and Mission," New York Times, March 31, 1980, p. A22; and "Soviet Nuclear Edge in Mid-80's is Envisioned by U.S. Intelligence," New York Times, May 13, 1980, p. A12. Some day scholars may investigate, systematically, the manipulation of somewhat distant threat projections for the purpose of influencing current debate.

28. Many commentators who have long histories of underestimating the Soviet threat suddenly, in 1980, when it suited their need to find arguments to oppose MX/MPS, were willing uncritically to project Soviet deployment of 20,000 or more silo-kill-capable warheads for the end of the 1980s. The general public should understand that the CIA's record of wisdom in strategic threat prediction is not particularly impressive. Moreover, the public should also appreciate that intelligence estimating, like arms control verification, has become a highly politicized activity. See Les Aspin, "Debate over U.S. Strategic Forecasts: A Mixed Record," Strategic Review 8, no. 3 (Summer 1980): 29-43; William T. Lee, "Debate over U.S. Strategic Forecasts: A Poor Record," ibid., pp. 44-57; and Aspin's "Rebuttal," ibid., pp. 57-59.

29. See William J. Perry, The Department of Defense Statement on the MX System and Ballistic Missile Defense, made before the Subcommittee on Research and Development of the Committee on Armed Services of the U.S. Senate, 96th Cong., 2nd sess. (Washington, D.C.: Department of Defense, March 12, 1980).

30. The case for LoADS deployment synergistically with MPS-based ICBMs is made in Carnes Lord, "The ABM Question," Commentary 69, no. 5 (May 1980): 31-38; and Gray, Ballistic Missile

Defense: A New Debate for a New Decade.

31. On the status of LoADS technology, see "Demonstration Planned for MX Defense System," Aviation Week and Space Technology 112, no. 24 (June 16, 1980): 220-21.

CHAPTER 6

1. This author offered a summary of his analysis in "What the MX Would Do to Soviet Defense Planners," letter to the editor, New York Times, April 16, 1980, p. A26.

2. See "Ballistic Missile Defense Tests Set," Aviation Week and Space Technology 112, no. 24 (June 16, 1980): 213-18.

3. At some risk of restating the obvious, it is sensible to assume that the Soviets, too, have a mirror-image problem (though they may be less well placed to recognize and compensate for the deficiency than is the United States).

4. See, for an important example, Anthony Austin, "Moscow Expert Says U.S. Errs on Soviet War Aims," New York Times, August 25, 1980, p. A2. Lieutenant General Mikhail Milshtein, the "Moscow expert" quoted in the Austin report, quite properly— in Soviet perspective— rejects Western ideas of "limited" central war. Unfortunately, many Western strategic thinkers fail to understand that what they, and what General Milshtein, mean by "limited" central war is quite different. In the Soviet view it is nonsensical to talk of "limited nuclear war"— nuclear war is waged for a total political goal in a militarily intelligent fashion. Military intelligence, in this context, means that one does not waste warheads on targets that are irrelevant to the security of the U.S.S.R. (that is, on many U.S. cities).

5. However, the major arguments in Joseph D. Douglass, Jr., and Amoretta Hoeber, Soviet Strategy for Nuclear War (Stanford, Calif.: Hoover Institution Press, 1979), are compatible with what this author understands, from all sources.

6. The target set may be derived from, inter alia, a careful reading of William F. Scott and Harriet F. Scott, The Soviet Control Structure, Final Report, SPC 575 (Arlington, Va.: System Planning Corp., 1980). The value, and the difficulty, in attacking the Soviet political control structure is outlined in Colin S. Gray, "Targeting Problems for Central War," Naval War College Review 33, no. 1 (January-February 1980): 3-21.

7. Carter administration spokesmen, correctly, have argued that Minuteman vulnerability is not the fault of the SALT process. But those spokesmen have neglected to acknowledge that back in 1972, SALT was "sold" in the United States, in part, on the basis of the claim that the limits on ICBM launcher numbers and, less directly,

on throw weight, in the interim agreement would help constrain the evolution of the Soviet hard-target kill threat.

8. See Clarence A. Robinson, Jr., "Soviets Boost ICBM Accuracy," Aviation Week and Space Technology 108, no. 14 (April 3, 1978): 14-16.

9. See David S. Sullivan, Soviet SALT Deception (Boston, Va.: Coalition for Peace Through Strength, 1979), pp. 1-3.

10. It is easy to make light of the engineering issues raised by the proposal to rebase Minuteman III. However, to select but a few, such a program would require a new canister, a new shock isolation system (missile to canister, canister to shelter), a new transporter/emplacer; probably a new C^3 system; a flight-test program to check on the solutions designed for the multiple technology interaction, and so on.

11. Unless the president were willing to press Congress very hard for a legislative waiver, the environmental impact statement requirement pertaining to Minuteman rebasing would impose a delay of two years on the start of new construction. Also, there is a legal-political problem: Representative Virginia Smith of Nebraska succeeded in appending an amendment to the Department of Defense Supplemental Appropriations Act of 1979, to the effect that MX must be based on the least productive agricultural land available. This should not affect Minuteman rebasing— but what about a Minuteman rebasing scheme explicitly, and honestly, defended as a "quick fix" pending MX availability?

12. Some proponents of Minuteman rebasing argue that a friendly local population of farmers constitutes, de facto, a powerful unofficial security force.

CHAPTER 7

1. R. James Woolsey, "Getting the MX Moving," Washington Post, May 22, 1980, p. A17.

2. As in Art. IV, para. 10, and Art. V of the SALT II Treaty. See U.S. Department of State, SALT II Agreement, Selected Documents no. 12A, Department of State Publication 8984 (Washington, D.C.: U.S. Government Printing Office, 1979).

3. See Kosta Tsipis, "The Accuracy of Strategic Missiles," Scientific American 233, no. 1 (July 1975): 14-23; and Bernard T. Feld and Kosta Tsipis, "Land-Based Intercontinental Ballistic Missiles," Scientific American 241, no. 5 (November 1979): esp. 54.

4. Well above the supposed 5 percent per annum "real" defense outlay increase currently projected for the next five years. In common with his Hudson Institute colleague Herman Kahn, this author

believes that a genuine U.S. "mobilization response" is a distinct possibility in the 1980s. He also believes that the failure of the SALT process would— as today— be a symptom of the fallout from a very general deterioration in U.S.-Soviet relations, not a principle cause of such deterioration. SALT reflects— it does not, by and large, mold— the political world.

5. By way of analogy, it is difficult even to conceive of the beginning of a serious U.S. SALT bargaining position with regard to air defenses and civil defense. In principle, both might be "negotiable"— but at what price?

6. A thesis propounded and defended by this author in Arms Control and European Security: Some Basic Issues, HI-3157/2-P (Croton-on-Hudson, N.Y.: Hudson Institute, 1980). Also see Daniel O. Graham, Shall America Be Defended: SALT II and Beyond (New Rochelle, N.Y.: Arlington House, 1979); and Edward N. Luttwak, "Why Arms Control Has Failed," Commentary 65, no. 1 (January 1978): 19–28.

7. Of course, it is always possible that the Soviet Union will respond to MX in a somewhat mindlessly competitive fashion, regardless of the quality of the U.S. determination to maintain the invulnerability of MX. For pessimistic analyses see William Kincade, "Will MX Backfire?" Foreign Policy no. 37 (Winter 1979–80): 43–58; and "MX: The Missile We Don't Need," Defense Monitor 8, no. 9 (October 1979): 5.

8. See Robert Legvold, "Strategic 'Doctrine' and SALT: Soviet and American Views," Survival 21, no. 1 (January/February 1979): 8–13.

9. It is possible that the abbreviated SALT II debate of 1979 was really the end of an era in arms control. The arms control inactivity of 1980, instead of constituting a hiatus between periods of SALT negotiating "business as usual," may be the beginning of a period wherein formal arms control processes are severely downgraded (in contrast with the 1970s). With the dust having largely settled on still-current arms control issues, a number of commentators—of different doctrinal persuasions—have begun to reassess the entire arms control enterprise. See Richard Burt, "International Security and the Relevance of Arms Control," Daedalus 110, no. 1 (Winter 1980): 159–77; and Colin S. Gray and Donald G. Brennan, Common Interests and Arms Control, HI-3218-P (Croton-on-Hudson, N.Y.: Hudson Institute, 1980).

10. Some well-informed commentators, early in 1979, believed that President Carter could not have SALT II without MX. See Clarence A. Robinson, Jr., "SALT 2 Approval Hinges on MX," Aviation Week and Space Technology 110, no 20 (May 14, 1979): 14–16; and Robin Ranger, "SALT II's Political Failure: The U.S. Senate Debate," RUSI Journal 125, no. 2 (June 1980): 50–51.

11. U.S. Department of State, SALT II Agreement, p. 27 (art. II).

12. See the excellent presentation in Paul K. Davis, Land-Mobile ICBM's, ACIS Working Paper no. 18 (Los Angeles: Center for International and Strategic Affairs, UCLA, 1980).

13. Representative Carr was wrong when he claimed: "MAP appears to present insurmountable verification problems. I do not see how a system can be at the same time deceptive and verifiable." U.S. House of Representatives, Committee on Armed Services, Panel on the Strategic Arms Limitation Talks and the Comprehensive Test Ban Treaty of the Intelligence and Military Application of Nuclear Energy Subcommittee, SALT II: An Interim Assessment, Report, 95th Cong., 2nd sess. (Washington, D.C.: U.S. Government Printing Office, 1978), "Dissenting Views," p. 39. The SALT verification factor has been considered very explicitly in the design and proposed operating practices concerning every major land-mobile ICBM basing alternative. Whatever else may be said against the system designers of MX/MPS, a cavalier attitude toward potential verification problems is not among them.

14. "Competitive performance" refers here to the global engagement of Soviet and American interests, not only to strategic arms activity.

15. The fact was not recognized in SALT One: Compliance/SALT Two: Verification, Selected Documents no. 7 (Washington, D.C.: Department of State, 1978). On SALT compliance issues see David S. Sullivan: "The Legacy of SALT I: Soviet Deception and U.S. Retreat," Strategic Review 7, no. 1 (Winter 1979): 26–41; and Soviet SALT Deception (Boston, Va.: Coalition for Peace Through Strength, 1979). For an overall analysis of this subject, see Amron H. Katz, Verification and SALT: The State of the Art and the Art of the State (Washington, D.C.: Heritage Foundation, 1979).

16. In response to which see U.S. Air Force, MX: Milestone II Draft Environmental Impact Statement, 5 vols. (Norton AFB, San Bernardino, Calif.: Space and Missile Systems Organization, U.S. Air Force, 1978); and Preliminary Draft Environmental Impact Statement for MX Deployment Area Selection and Land Withdrawal/Acquisition, prepared for USAF, Ballistic Missile Office, Norton AFB, Calif., 9 vols. (Santa Barbara, Calif.: HDR Sciences, 1980).

17. See the testimony of Air Force Chief of Staff Lew Allen and Secretary of Defense Harold Brown in, respectively, U.S. Senate, Committee on Foreign Relations, The SALT II Treaty, Hearings, 96th Cong., 1st sess. (Washington, D.C.: U.S. Government Printing Office, 1979, pt. 1, p. 424, and pt. 2, pp. 245–46.

18. In terms of military "style" it has not been "the Soviet way" to invest heavily in deceptive instruments. Just as the Soviets have been unfriendly to the idea of decoys (why waste rubles, space, and

weight on a decoy, when one could be doing serious things—such as building real warheads?), the idea of constructing large numbers of empty silos may not find favor.

19. See Jeffrey T. Richelson, "Multiple Aim Point Basing: Vulnerability and Verification Problems," Journal of Conflict Resolution 23, no. 4 (December 1979): 613-28; and Stephen M. Meyer, "Verification and the ICBM Shell-Game," International Security 4, no. 2 (Fall 1979): 40-68.

20. A "worst case," from the perspective of verification, would be deployment of relatively small ICBMs, with decoys, using area rather than point security in an area that had many public roads.

21. See the excellent discussion in Davis, Land-Mobile ICBM's.

22. Harold Brown, speech at the Naval War College, Newport, R.I., August 20, 1980.

23. For example, see McGeorge Bundy, "To Cap the Volcano," Foreign Affairs 48, no. 1 (October 1969): 1-20, for a contemporary celebration of the U.S. abandonment of the concept of strategic superiority.

24. For useful reminders of those objectives, see Donald G. Brennan, "Setting and Goals of Arms Control," in Donald G. Brennan, ed., Arms Control, Disarmament, and National Security (New York: Braziller, 1961), pp. 19-42; and Thomas C. Schelling and Morton H. Halperin, Strategy and Arms Control (New York: Twentieth Century Fund, 1961).

25. In his prepared statement on SALT II issues, my late colleague Donald G. Brennan attempted to direct attention to these fundamental matters. See U.S. Senate, Committee on Foreign Relations, The SALT II Treaty, pt. 4, pp. 369-75.

26. Jeremy Stone, Strategic Persuasion: Arms Limitation Through Dialogue (New York: Columbia University Press, 1967).

27. See the careful, almost too careful (so painfully "balanced" are the judgments), analysis in Thomas W. Wolfe, The SALT Experience (Cambridge, Mass.: Ballinger, 1979), esp. pp. 247-50.

28. This author would like to be able to cite authoritative analyses of the Soviet approach to SALT, or of the Soviet approach to arms control in general. However, strange though it may seem, the often intensive East-West arms control interactions of the 1970s have yet to spawn a literature that has even the appearance of enduring merit. The best single recent study is Thomas Wolfe, The SALT Experience. Wolfe's work, though always worthy of respect for its meticulous scholarship, can be frustrating because it typically declines to make judgments. See Samuel B. Payne, Jr., The Soviet Union and SALT (Cambridge, Mass.: MIT Press, 1980).

29. See Kincade, "Will MX Backfire?"

30. Estimates of psi resistance necessarily are somewhat conjectural, but the most modern Soviet silos appear to be hardened to resist 6,200 psi.

31. See Colin S. Gray, The Future of Land-Based Missile Forces, Adelphi Papers no. 140 (London: International Institute for Strategic Studies, 1977), pp. 17-18.

32. "Serious Trouble Develops for Arms Treaty with Soviets," New York Times, September 28, 1979, p. A2.

33. These themes permeate Colin S. Gray, Strategy and the MX, Critical Issues series (Washington, D.C.: Heritage Foundation, 1980).

34. See the prepared statement of Richard Pipes in U.S. House of Representatives, Permanent Select Committee on Intelligence, Subcommittee on Oversight, Soviet Strategic Forces, Hearings, 96th Cong., 2nd sess. (Washington, D.C.: U.S. Government Printing Office, 1980), pp. 2-7; and Lawrence Freedman, U.S. Intelligence and the Soviet Strategic Threat (London: Macmillan, 1977), pp. 197-98.

35. P.D. 59 may be hailed by conservative optimists as the arrival of the strategy millennium, but there are some historic grounds for skepticism. Specifically, high-level policy decisions in favor of a militarily intelligent set of war plans have been issued twice before— in the early 1960s by Robert McNamara, and by James Schlesinger following the approval of NSDM 242 in April 1974—but, somehow, a strategy worthy of the name failed to be translated into SIOP planning. P.D. 59 may fare better, but the precedents are discouraging. I am grateful to my colleague Paul Bracken for impressing upon me the sad fate, at the all-important levels of SIOP design implementation and military hardware and software acquisition, of all pre-P.D. 59 thrusts toward a sensible strategy.

36. Department of Defense Annual Report, Fiscal Year 1980 (Washington, D.C.: U.S. Government Printing Office, 1979), p. 118.

37. Since SALT began in November 1969, first with reference to the 3-MIRVed SS-9 Mod-4, and then with reference to the MIRVed fourth-generation ICBMs (SS-17s, -18s, and -19s), the United States has given the (accurate) impression of negotiating "under the gun" of an impending, intolerable Soviet hard-target counterforce threat.

38. It is interesting to note that, learning nothing from history, some commentators argued, in the fall of 1980, that the U.S. should hasten to ratify the SALT II Treaty, as negotiated and signed in June 1979, because the military trend is not advantageous (to the United States) and, therefore, one should not anticipate being able to renegotiate SALT II for an improved outcome. See "Expert [Marshall D. Shulman] on Soviets Expects Trouble," New York Times, August 30, 1980, p. 5.

39. The standard official reply to the charge that silo-housed Minuteman should not have been permitted to become vulnerable, is that "that is why we maintain a strategic-forces triad." In his speech at the Naval War College, Newport, R.I., on August 20, 1980, explaining P.D. 59, Defense Secretary Harold Brown said that "[t]he other [than ICBM] elements of our strategic force—each of which will be improving rapidly in the early 1980's—enable us to maintain the balance and a survivable deterrent during this temporary vulnerability of ICBM's."

CHAPTER 8

1. Particularly thorough was the set of ICBM survivability studies conducted at the Institute for Defense Analysis in 1966-67, known as "Strat-X." See Lawrence Freedman, U.S. Intelligence and the Soviet Strategic Threat (London: Macmillan, 1977), pp. 121-22.

2. Henry Kissinger, American Foreign Policy: Three Essays (London: Weidenfeld and Nicholson, 1969), p. 21.

3. For example, see Tom Wicker, "After the MX, What?" New York Times, March 25, 1980.

4. A careful and persuasive analysis of the importance of MX is Richard Burt, "Reassessing the Strategic Balance," International Security 5, no. 1 (Summer 1980): 43-46.

5. For example, in Colin S. Gray, Strategy and the MX, Critical Issues series (Washington, D.C.: Heritage Foundation, 1980).

6. For a careful and not unfriendly study of launch under attack, see Richard Garwin, "Launch Under Attack to Redress Minuteman Vulnerability?" International Security 4, no. 3 (Winter 1979/80): 117-39.

7. See Colin S. Gray, The Future of Land-Based Missile Forces, Adelphi Papers no. 140 (London: International Institute for Strategic Studies, 1977), pp. 8-28. Also see Desmond Ball, "The MX Basing Decision," Survival 22, no. 2 (March/April 1980): 58-59.

8. For example, see the prepared testimony of Richard L. Garwin to the Committee on Armed Services of the U.S. House of Representatives, February 7, 1979, pp. 29-39 (mimeo). SUMS has been advocated by, inter alia, Senator Mark Hatfield. See his articles "SUM Strategy," Armed Forces Journal International, January 1980, p. 35; and "SUM: It Adds Up," ibid., February 1980, p. 66.

9. For example, see Bill Keller, "Attack of the Atomic Tidal Wave: Sighted S.U.M., Sank Same," Washington Monthly, May 1980, pp. 53-58.

10. A useful critique of SUM is Jake Garn, "SUM: Simplistic, Unworkable and Maladroit," Armed Forces Journal International, May 1980, pp. 55-56, 65.

11. Air-mobile variants for ICBM deployment were revived in popularity in 1978 (having been shelved since 1976), largely as a consequence of the so-called Press Report (after Dr. Frank Press, the president's science adviser), which claimed to have found a number of basic flaws in the leading land-mobile/movable basing concepts. See Jeffrey M. Lenorovitz, "Air Force Restudying MX Basing Plan," Aviation Week and Space Technology 110, no. 3 (January 15, 1979): 21. Also see "MX Still Zigzagging," Air Force Magazine 62, no. 2 (February 1979): 14.

12. Furthermore, one wonders about the "environmental impact" of potential crisis-time deployment of such systems on a set of "austere" airfields that could, in principle, number in the thousands. See Ball, "The MX Basing Decision," p. 59.

13. See Richard Burt, "Reassessing the Strategic Balance," International Security 5, no. 1 (Summer 1980): 45.

14. See Rowland Evans and Robert Novak, "Unheeded Warnings About the ICBMs," Washington Post, January 12, 1979.

15. Harold Brown made a reasonably strong statement to this effect in Department of Defense Annual Report, Fiscal Year 1980 (Washington, D.C.: U.S. Government Printing Office, 1979), p. 118.

16. U.S. Department of State, SALT II Agreement, Selected Document no. 12A, Dept. of State publication 8984 (Washington, D.C.: U.S. Government Printing Office, 1979), p. 50 (para. 3). At Mr. Carter's instigation, studies were conducted within the U.S. government on the implications of strategic-force level reductions to a level as low as 250 launchers (though Mr. Carter may have been more interested in 250 warheads, or 25 MX ICBMs). On the subject of deep reductions in force levels, see Colin S. Gray and Keith B. Payne, U.S. Strategic Posture and Deep Reductions in Force Levels Through SALT, HI-3195-RR (Croton-on-Hudson, N.Y.: Hudson Institute, 1980).

17. For the foreseeable future only an ICBM force can execute the prompt hard-target counterforce mission with any high degree of assurance of success. However, as John Steinbruner has argued, from the perspective of "command stability" the SSBN force is far less desirable than are ICBMs: "National Security and the Concept of Strategic Stability," Journal of Conflict Resolution 22, no. 3 (September 1978): 422.

18. An ALCM force could fail catastrophically in mission performance in ways that an MPS-based ICBM force could not (that is, the ALCM force might not receive the tactical warning necessary to reach its safe escape distance from runways, and Soviet multilayered air defenses might impose a truly dramatic measure of loss).

19. Thereby contributing to stability in the escalation process by enforcing a tolerable (for the United States) post (first and second counterforce) strike balance.

20. See Harold Brown's comparative checklist in Department of Defense Annual Report, Fiscal Year 1980, p. 118.

21. See Paul Wolfowitz, "The Proposal to Launch on Warning," in U.S. Senate, Committee on Armed Services, Authorization for Military Procurement and Research and Development, Fiscal Year 1971, and Reserve Strength, Hearings, pt. 3, 91st Cong., 2nd sess. (Washington, D.C.: U.S. Government Printing Office, 1970), pp. 2278-82.

22. For a thorough technical review, see Garwin, "Launch Under Attack to Redress Minuteman Vulnerability?"

23. Nor would the United States be likely to know whether the incoming Soviet missile attack was directed at the ICBM fields.

24. Such a decision has not yet been taken. "However, they [the Soviets] would have to consider the possibility of our having launched the MINUTEMAN force before their ICBMs arrived, even though we have not made 'launch under the attack' a matter of policy for a very good reason. ..." Brown, Department of Defense Annual Report, Fiscal Year 1980, p. 15.

25. A good case can be made for the value of studying just how one would execute SIOP-level attack options in an LUA context. While granting the extreme undesirability of LUA as the U.S. firing tactic, some sophisticated analysis of what the U.S. might prefer to launch against, and why, could be enlightening.

26. Paul K. Davis, Land-Mobile ICBMs: Verification and Breakout, ACIS Working Paper no. 18 (Los Angeles: Center for International and Strategic Affairs, UCLA, 1980).

27. This is not to assert a simpleminded action-reaction model of the arms competition. The author has assailed such thinking in "The Arms Race Phenomenon," World Politics 24, no. 1 (October 1971): 39-79; and in The Soviet-American Arms Race (Farnborough, England: Saxon House/D. C. Heath, 1976), esp. chs. 2 and 4. The U.S. defense debate over Soviet reaction processes—in the current case, to MX/MPS—is in urgent need of the wisdom that should be obtainable from the multivolume classified history of the arms race that the director of new assessment in the Office of the Secretary of Defense commissioned many years ago. This critically important, if not definitive, study may answer many of the questions that continue to tease students of Soviet-American arms race dynamics.

28. Also, it would help to know why one is competing and to understand the character of the competition and the requirements of deterrence considered in broad, as well as narrow, perspective. For a rare and valuable analysis of the Soviet-American strategic competitive relationship, see Bernard S. Albert, "The Strategic Arms Competition with the U.S.S.R.—What Is It and How Are We Doing?" Comparative Strategy 1, no. 3 (1979): 139-67.

29. There is a very important sense in which MIRV was deployed "because it was ready"—which is not to say that MIRV technology lacked for strategic rationales. For an instructive history of MIRV, see Ted Greenwood, Making the MIRV: A Study of Defense Decision Making (Cambridge, Mass.: Ballinger, 1975).

30. Since the Minuteman III production line was closed in 1979, anyone considering extending the service life of that missile—without reopening the production line—has to consider the fact that the Air Force will be able to find spare parts for the weapon only from the existing inventory of spares, and by cannabilizing "off-line" missiles.

31. See Harold Brown, Department of Defense Annual Report, Fiscal Year 1981 (Washington, D.C.: U.S. Government Printing Office, 1980), p. 133; and Clarence A. Robinson, Jr., "Reagan Details Defense Boost," Aviation Week and Space Technology 113, no. 19 (November 10, 1980): 16.

32. For example, while the United States destroyed its more than 1,500-strong B-47 medium jet bomber force after it was retired in the early 1960s (a total of 2,042 B-47s were built), the Soviet Union, in 1980, continues to fly its B-47 equivalent, the Tu-16 Badger (roughly 2,000 were built, with first production in 1952 and squadron service beginning in 1955).

33. Some weaponized laser proponents today have much of the fervor that once characterized true believers in "air power" and deep-striking, independently operating armored forces. As it happened, those prophets of the 1920s and 1930s were shown to have been probably half correct. However, there is always the possibility that a band of true believers, in love with a particular technology, might prove to be totally correct: not for very long, perhaps, but long enough for the next war to be lost should their advice not be heeded.

34. For example, the tank might have been the war winner of World War I—the reasons why it was not merit close attention.

CHAPTER 9

1. For an interesting recent assault on this subject, from the quantified social science view, see Theresa C. Smith, "Arms Race Instability and War," Journal of Conflict Resolution 24, no. 2 (June 1980): 253-84. See "Kissinger's Critique," The Economist, February 3, 1979, p. 22, for a terse, skeptical, and persuasive dismissal of the putative arms race-war connection.

2. Henry S. Rowen, "Formulating Strategic Doctrine," in Commission on the Organization for the Conduct of Foreign Policy, vol. 4, app. K, Adequacy of Current Organization: Defense and Arms Control (Washington, D.C.: U.S. Government Printing Office, 1975), p. 227.

3. Ibid., p. 228.

4. Ibid., pp. 231-32. Nonetheless, Mr. McNamara's policy failure was noticeably personal. Rowen goes on to argue that "[t]he implementation of Secretary McNamara's flexible options initiative in the early 1960's was aborted in large measure by the withdrawal of his interest and support" (p. 232).

5. All three of these points may be found in Harold Brown, speech at the Naval War College, Newport, R.I., August 20, 1980.

6. A classic "period piece" was Daedalus 104, no. 3 (Summer 1975), with the general title "Arms, Defense Policy and Arms Control." Readers interested in seeing how times have changed are invited to compare the contents of that issue of Daedalus with the contents of its 1980 equivalent, "U.S. Defense Policy in the 1980s" (Fall and Winter issues).

7. Outstanding works were Ted Greenwood, Making the MIRV (Cambridge, Mass.: Ballinger, 1975); Edmund Beard, Developing the ICBM: A Study in Bureaucratic Politics (New York: Columbia University Press, 1976); and Harvey M. Sapolsky, The Polaris System Development: Bureaucratic and Programmatic Success in Government (Cambridge, Mass.: Harvard University Press, 1972). The study that "showed the way" was Graham T. Allison, Essence of Decision: Explaining the Cuban Missile Crisis (Boston: Little, Brown, 1971).

8. For example, in Colin S. Gray, "The Arms Race Phenomenon," World Politics 24, no. 1 (October 1971): 39-79.

9. Most aggressors would prefer that their victims choose not to resist.

10. Testimony in U.S. Senate, Committee on Armed Services, Preparedness Investigating Subcommittee, Status of U.S. Strategic Power, Hearings, pt. 1, 90th Cong., 2nd sess. (Washington, D.C.: U.S. Government Printing Office, 1969), pp. 117-18.

11. See Bernard Brodie, Sea Power in the Machine Age (Princeton, N.J.: Princeton University Press, 1941), p. 254.

12. Samuel Huntington, "Arms Races: Prerequisites and Results," in Carl J. Friedrich and Seymour E. Harris, eds., Public Policy, 1958 (Cambridge, Mass.: Graduate School of Public Administration, Harvard University, 1958), pp. 65-75.

13. Through much of the 1960s, and lingering on into the 1970s, it was fairly orthodox to maintain that arms control could be pursued more effectively in a tacit than in a formal mode. The new arms control thinking of the very late 1950s and the early 1960s was in part a reaction against what were viewed as fruitless diplomatic spectacles. It was believed that there could, and should, be a permanent arms control dialogue between the superpowers—a process to which very prominent diplomatic "events" would have very little to contribute.

14. See Albert Wohlstetter, Legends of the Arms Race (Washington, D.C.: U.S. Strategic Institute, 1975); and Colin S. Gray, The Soviet-American Arms Race (Farnborough, England: Saxon House/D. C. Heath, 1976).

15. This argument can be overdrawn. The point to be registered is that it is virtually inconceivable that the Soviets would cancel a program in actual, and prospective, support of which "a cast of thousands had been assembled." (For example, there would be no Soviet B-1 decision à la Carter in 1977.)

16. A. J. Wohlstetter et al., Selection and Use of Strategic Air Bases, R-266 (Santa Monica, Calif.: RAND, 1954).

17. See U.S. House of Representatives, Committee on Armed Services, Panel on the Strategic Arms Limitation Talks and Comprehensive Test Ban Treaty of the Intelligence and Military Application of Nuclear Energy Subcommittee, SALT II: An Interim Assessment, Report, 95th Cong., 2nd sess. (Washington, D.C.: U.S. Government Printing Office, 1978), pp. 45-46. Carr's analysis is dubious on technical cost-effective grounds, and may address the wrong question. A point that he should have made is that it is far from certain just how important believed cost-exchange ratios are. Because the United States could almost certainly proliferate MPSs more cheaply than the Soviets could proliferate warheads with the appropriate characteristics, it does not follow that the Soviets would necessarily thereby be discouraged from engaging in a warhead versus MPS competition.

18. Such a targeting rule would not lead to a strictly "efficient" allocation of hard-target kill-capable resources, but the insurance argument should overwhelm any optimism that would rest upon high single shot kill probabilities derived from limited and "artificial" test data.

19. For example, in the 1950s the Soviet Union procured a vast inventory of day air-defense interceptors—quite useless if SAC had paid an unfriendly visit at night or in bad weather.

20. On Soviet defense decision making, see Arthur J. Alexander, Decision-Making in Soviet Weapons Procurement, Adelphi Papers nos. 147-48 (London: International Institute for Strategic Studies, Winter 1978/79); David Holloway: "Decision-Making in Soviet Defence Policies," in Prospects of Soviet Power in the 1980's, Pt. 2, Adelphi Papers no. 152 (London: International Institute for Strategic Studies, 1979), pp. 24-31; and "Military Power and Political Purpose in Soviet Policy," Daedalus 109, no. 4 (Fall 1980): 13-30.

21. See John Erickson, "The Soviet Military System: Doctrine, Technology and 'Style,'" in John Erickson and E. J. Feuchtwanger, eds., Soviet Military Power and Performance (Hamden, Conn.: Archon, 1979), pp. 28-31.

22. The absence of U.S. homeland defenses may be interpreted by some Soviet analysts as evidence of U.S. confidence in the total efficacy of its first strike (improbable though this may appear).

23. Although it could be the action of a state prepared to add insurance upon insurance—and where more appropriate to superinsure than with respect to the ICBM force?

24. There have been many very general Soviet statements to the effect that the U.S.S.R. "will not stand idly by" in the event that MX/MPS comes to pass, or "this will not go unanswered," and so on. However, to date there has been no specific, public Soviet commitment to any particular theme of response.

25. It is not only Americans who have downplayed the significance of the vulnerable ICBM problem. The director of the International Institute for Strategic Studies, Christoph Bertram, has given as his opinion that "[t]hough no doubt a problem [that Minuteman vulnerability might open a "window of opportunity" for the Soviet Union to exploit] which strategic planners would have to take into account, this did not seem to merit quite the central attention it received in the debate." Strategic Survey, 1979 (London: International Institute for Strategic Studies, 1980), p. 5.

CHAPTER 10

1. See Thomas Schelling: "Surprise Attack and Disarmament," in Klaus Knorr, ed., NATO and American Security (Princeton, N.J.: Princeton University Press, 1959), ch. 8; "Reciprocal Measures for Arms Stabilization," in Donald G. Brennan, ed., Arms Control, Disarmament, and National Security (New York: Braziller, 1961), ch. 9; The Strategy of Conflict (Cambridge, Mass.: Harvard University Press, 1960), chs. 9-10; and Arms and Influence (New Haven: Yale University Press, 1966), ch. 6.

2. See L. L. Farrar, Jr., The Short-War Illusion: German Policy, Strategy and Domestic Affairs, August-December 1914 (Santa Barbara, Calif.: ABC-Clio, 1973), ch. 1.

3. Carl von Clausewitz, On War, Michael Howard and Peter Paret, eds. (Princeton, N.J.: Princeton University Press, 1978), p. 605.

4. Civilian strategic theorists, as Bernard Brodie has observed, have tended to be very neglectful of the "after deterrence, what?" question. Bernard Brodie, "The Development of Nuclear Strategy," International Security 2, no. 4 (Spring 1978): 66.

5. Basil Liddell Hart, Strategy: The Indirect Approach (London: Faber and Faber, 1967), p. 366.

6. In CIA, Soviet Civil Defense (Washington, D.C.: CIA, 1978), p. 4.

7. See Gray, "Nuclear Strategy: The Case for a Theory of Victory," International Security 4, no. 1 (Summer 1979): 54-87.

8. I address this, and related questions, in my article "Strategic Stability Reconsidered," Daedalus 109, no. 4 (Fall 1980): 135-54.

9. It is just (barely) possible to argue that there is a lag in Soviet defense analysis. In short, the Soviets tend to appreciate some of the more important analytical complexities perhaps five years after the U.S. defense community has attained a substantial understanding of them. This author does not subscribe to this theory, though it is possible—on occasion—that it may have some limited merit. Readers are recommended to consider the arguments in Fritz Ermarth, "Contrasts in American and Soviet Strategic Thought," International Security 3, no. 2 (fall 1978): 138-55. It is possible that Soviet defense analysts have yet to consider in detail the likely structure and course of an arms competition in the late 1980s wherein both sides might be deploying ICBMs in MPS complexes, or other nonfixed site modes, and some hard-point BMD.

10. The Strategic Arms Limitation Talks, Special Report no. 46 (Washington, D.C.: Department of State, 1978), p. 3.

11. Intimations of a change in U.S. official (Defense Department, at least, if not the State Department) thinking on hard-target counterforce were first discernible in Harold Brown, Department of Defense Annual Report, Fiscal Year 1980 (Washington, D.C.: U.S. Government Printing Office, 1979): 78—"...[A]ttacks on these targets [missile silos, command bunkers, and nuclear weapon storage sites] would not disarm an enemy in a first-strike (because of his survivable non-ICBM forces), but on a second-strike could suppress his withheld missiles and recycling bombers that could otherwise be used against vital targets." P.D. 59 of July 25, 1980 showed a very marked evolution of official thinking. See Desmond Ball, Developments in U.S. Strategic Nuclear Policy Under the Carter Administration, ACIS Working Paper no. 21 (Los Angeles: Center for International and Strategic Affairs, UCLA, 1980).

12. If the United States is leading the process of escalation, as seems plausible to predict, the higher levels of that process could comprise, in essence, a situation where Soviet society (and particularly the essential assets of the Soviet state—in Soviet estimation) would be dramatically less vulnerable to damage than would the United States. If this asymmetry is real, or is believed in Moscow to be real, the Soviet Union has a theory of escalation dominance. This author is not willing to predict in detail just how well the Soviets might do in protecting their homeland, but he does wish that the U.S. defense

community would think through which U.S. war-waging strategy might, most economically, effect a definitive Soviet defeat. The new and evolving richness of the U.S. targeting menu, at least as reported, continues to evade such questions as what victory against the Soviet Union would mean—and how that goal might be forwarded. MX/MPS should help greatly in the control of conflict, but behind that system there needs to be an overarching theory of the conduct of war at tolerable cost. See Colin S. Gray and Keith B. Payne, "Victory Is Possible," Foreign Policy, no. 39 (Summer 1980): 14-27.

CHAPTER 11

1. For convenience, see "MX Associate Contractors Listed," Aviation Week and Space Technology 113, no. 2 (July 14, 1980): 70.

2. Colin S. Gray, The Soviet-American Arms Race (Farnborough, England: Saxon House/D. C. Heath, 1976), pp. 47-53.

3. However, for one of the better compilations of writings on this subject, readers are referred to Stephen Rosen, ed., Testing the Theory of the Military-Industrial Complex (Lexington, Mass.: Lexington Books, 1973).

4. Christopher E. Paine, in Bulletin of the Atomic Scientists 36, no. 2 (February 1980): 12-16.

5. On a Soviet MIC, see Mikhail Agursky and Hannes Adomeit, "The Soviet Military-Industrial Complex," Survey 24, no. 2 (Spring 1979): 106-24.

6. An option strongly recommended in Stephen Rosen, "Safeguarding Deterrence," Foreign Policy no. 35 (Summer 1979): 109-23.

7. On the basis of poll evidence seen by this author, and perhaps surprisingly, strong local objections tend not to stress the potential target aspect of the local situation.

8. Although the strategic argument against split basing is unambiguously negative, there is no question that the Air Force could live with the postattack problem of status update. Instead of the entire MX ICBM force intercommunicating for retargeting, that exercise would be conducted within each split basing area.

9. See U.S. Air Force, MX: Milestone II Draft Environmental Impact Statement, 5 vols. (Norton AFB, San Bernardino, Calif.: Space and Missile Systems Organization, U.S. Air Force, 1978); and Preliminary Draft Environmental Impact Statement for MX Deployment Area Selection and Land Withdrawal/Acquisition, prepared for USAF, Ballistic Missile Office, Norton AFB, Calif., 9 vols. (Santa Barbara, Calif.: HDR Sciences, 1980).

CHAPTER 12

    1. The most informative commentary on the British strategic forces' choice is Peter Nailor and Jonathan Alford, <u>The Future of Britain's Deterrent Force</u>, Adelphi Papers no. 156 (London: International Institute for Strategic Studies, 1980).

GLOSSARY OF ABBREVIATIONS

ABM — Antiballistic missile

ALCM — Air-launched cruise missile; to be deployed on 151 B-52s (20 per aircraft) beginning in 1982

ASW — Antisubmarine warfare

BMD — Ballistic missile defense; a more generic term than ABM (antiballistic missile)

CEP — Circular error probable. The measure of missile accuracy. (An estimate of the radius of a circle within which 50 percent of reentry vehicles are expected to land.)

C^3I — Command, control, communications, and intelligence

CONUS — Continental United States

ICBM — Intercontinental ballistic missile (missile with range greater than 3,500 nautical miles)

IOC/FOC — Initial operational capability; full operational capability

LoADS — Low-altitude defense system

LRCM — Long range cruise missile

MAD — Mutual assured destruction (a theory of deterrence)

MIRV — Multiple independently targetable reentry vehicle

MRBM/IRBM — Medium range ballistic missile (Missile with range of 430 to 1,300 nautical miles); intermediate range ballistic missile (missile with range of 1,300 to 3,500 nautical miles)

MX/MPS   The MX (missile, experimental) ICBM deployed in multiple protective structures; "baseline MX/MPS" refers to 200 MX ICBMs deployed in 4,600 MPS

R,D,T, and E   Research, development, test, and evaluation

RV   Reentry vehicle

SIOP   Single integrated operational plan (the U.S. nuclear war plan)

SLBM   Submarine-launched ballistic missile

SSBN   Nuclear-powered ballistic-missile-firing submarine

BIBLIOGRAPHY

BOOKS

Allison, Graham T. *Essence of Decision: Explaining the Cuban Missile Crisis.* Boston: Little, Brown, 1971.

Beard, Edmund. *Developing the ICBM: A Study in Bureaucratic Politics.* New York: Columbia University Press, 1976.

Brodie, Bernard. *Sea Power in the Machine Age.* Princeton, N.J.: Princeton University Press, 1941.

Chayes, Abram, and Jerome B. Wiesner, eds. *ABM: An Evaluation of the Decision to Deploy an Antiballistic Missile System.* New York: Harper and Row, 1969.

Clausewitz, Carl von. *On War.* Edited by Michael Howard and Peter Paret. Princeton, N.J.: Princeton University Press, 1978.

Collins, John M. *U.S.-Soviet Military Balance: Concepts and Capabilities, 1960-1980.* New York: McGraw-Hill, 1980.

Craig, Gordon A. *Germany, 1866-1945.* New York: Oxford University Press, 1978.

Douglass, Joseph D., Jr., and Amoretta M. Hoeber. *Soviet Strategy for Nuclear War.* Stanford, Calif.: Hoover Institution Press, 1979.

Erickson, John, and E. J. Feuchtwanger, eds. *Soviet Military Power and Performance.* Hamden, Conn.: Archon, 1979.

Farrar, L. L., Jr. *The Short-War Illusion: German Policy, Strategy and Domestic Affairs, August-December 1914.* Santa Barbara, Calif.: ABC-Clio, 1973.

Freedman, Lawrence. *U.S. Intelligence and the Soviet Strategic Threat.* London: Macmillan, 1977.

Gibbs, N. H. *Grand Strategy: volume I, Rearmament Policy.* London: Her Majesty's Stationery Office, 1976.

Goldhammer, Herbert. The Soviet Soldier: Soviet Military Management at the Troop Level. New York: Crane, Russak, 1975.

Graham, Daniel O. Shall America Be Defended: SALT II and Beyond. New Rochelle, N.Y.: Arlington House, 1979.

Gray, Colin S. The Soviet-American Arms Race. Farnborough, England: Saxon House/D. C. Heath, 1976.

Greenwood, Ted. Making the MIRV: A Study of Defense Decision Making. Cambridge, Mass.: Ballinger, 1975.

Hoeber, Francis. Slow to Take Offense: Bombers, Cruise Missiles, and Prudent Deterrence. Washington, D.C.: Center for Strategic Studies, Georgetown University, 1977.

Holst, Johan J., and William Schneider, Jr. Why ABM? Policy Issues in the Missile Defense Controversy. New York: Pergamon, 1969.

Kahan, Jerome H. Security in the Nuclear Age: Developing U.S. Strategic Arms Policy. Washington, D.C.: Brookings Institution, 1975.

Kissinger, Henry. American Foreign Policy: Three Essays. London: Weidenfeld and Nicholson, 1969.

———. White House Years. Boston: Little, Brown, 1979.

Liddell Hart, Basil. Strategy: The Indirect Approach. London: Faber and Faber, 1967.

Payne, Samuel B., Jr. The Soviet Union and SALT. Cambridge, Mass.: M.I.T. Press, 1980.

Quester, George H. Offense and Defense in the International System. New York: John Wiley, 1977.

Rosen, Stephen, ed. Testing the Theory of the Military-Industrial Complex. Lexington, Mass.: Lexington Books, 1973.

Sapolsky, Harvey M. The Polaris System Development: Bureaucratic and Programmatic Success in Government. Cambridge, Mass.: Harvard University Press, 1972.

Schelling, Thomas C. The Strategy of Conflict. Cambridge, Mass.: Harvard University Press, 1960.

———. Arms and Influence. New Haven: Yale University Press, 1966.

Schelling, Thomas C., and Morton H. Halperin. Strategy and Arms Control. New York: Twentieth Century Fund, 1961.

Spielmann, Karl F. Analyzing Soviet Strategic Arms Decisions. Boulder, Colo.: Westview, 1978.

Stone, Jeremy. Strategic Persuasion: Arms Limitation Through Dialogue. New York: Columbia University Press, 1967.

Stone, Norman. The Eastern Front, 1914-1917. London: Hodder and Stoughton, 1975.

Strategic Survey, 1979. London: International Institute for Strategic Studies, 1980.

Van Cleave, William R., and W. Scott Thompson, eds. Strategic Options for the Early Eighties: What Can Be Done? New York: National Strategy Information Center, 1979.

Wolfe, Thomas W. The SALT Experience. Cambridge, Mass.: Ballinger, 1979.

Zumwalt, Elmo. On Watch. New York: Quadrangle, 1976.

ARTICLES AND CHAPTERS IN BOOKS

Brennan, Donald G. "Setting and Goals of Arms Control." In Donald G. Brennan, ed., Arms Control, Disarmament, and National Security, pp. 19-42. New York: Braziller, 1961.

Huntington, Samuel P. "Arms Races: Prerequisites and Results." In Carl J. Friedrich and Seymour E. Harris, eds., Public Policy, 1958, pp. 41-86. Cambridge, Mass.: Graduate School of Public Administration, Harvard University, 1958.

Rowen, Henry S. "The Evolution of Strategic Nuclear Doctrine." In Laurence Martin, ed., Strategic Thought in the Nuclear Age, pp. 131-56. Baltimore: Johns Hopkins University Press, 1979.

Schelling, Thomas C. "Surprise Attack and Disarmament." In Klaus Knorr, ed., NATO and American Security, ch. 8. Princeton, N.J.: Princeton University Press, 1959.

Van Cleave, William R. "Quick Fixes to U.S. Strategic Nuclear For-
ces." In W. Scott Thompson, ed., National Security in the 1980's:
From Weakness to Strength, ch. 5. San Francisco: Institute for
Contemporary Studies, 1980.

JOURNAL ARTICLES AND OTHER SOURCES

Adelman, Kenneth L. "Rafshooning the Armageddon: Selling SALT."
Policy Review no. 9 (Summer 1979): 85-102.

Agursky, Mikhail, and Hannes Adomeit. "The Soviet Military-Industrial Complex." Survey 24, no. 2 (Spring 1979): 106-24.

Albert, Bernard S. "The Strategic Arms Competition with the U.S.S.R.—
What Is It and How Are We Doing?" Comparative Strategy 1, no.
3 (1979): 139-67.

Alexander, Arthur J. Decision-Making in Soviet Weapons Procurement. Adelphi Papers nos. 147-48. London: International Institute for Strategic Studies, Winter 1978/79.

"Arms, Defense Policy and Arms Control." Daedalus 104, no. 3
(Summer 1975).

Aspaturian, Vernon V. "Soviet Global Power and the Correlation of
Forces." Problems of Communism 29 (May-June 1980): 1-18.

Aspin, Les. "Debate over U.S. Strategic Forecasts: A Mixed Record."
Strategic Review 8, no. 3 (Summer 1980): 29-43.

——. "Rebuttal." Strategic Review 8, no. 3 (Summer 1980): 57-59.

Ball, Desmond. Deja Vu: The Return to Counterforce in the Nixon
Administration. Los Angeles: California Seminar on Arms Control
and Foreign Policy, 1974.

——. Developments in U.S. Strategic Nuclear Policy Under the Carter Administration. ACIS Working Paper no. 21. Los Angeles:
Center for International and Strategic Affairs, UCLA, 1980.

——. "The MX Basing Decision." Survival 22, no. 2 (March/April
1980): 58-65.

——. "Research Note: Soviet ICBM Deployment." Survival 22, no. 4 (July/August 1980): 167-70.

Brodie, Bernard. "On the Objectives of Arms Control." International Security 1, no. 1 (Summer 1976): 17-36.

——. "The Development of Nuclear Strategy." International Security 2, no. 4 (Spring 1978): 65-83.

Bundy, McGeorge. "To Cap the Volcano." Foreign Affairs 48, no. 1 (October 1969): 1-20.

——. "Maintaining Stable Deterrence." International Security 3, no. 3 (Winter 1978/79): 5-16.

——. "The Future of Strategic Deterrence." Survival 21, no. 6 (November/December 1979): 268-72.

Burt, Richard. "Reassessing the Strategic Balance." International Security 5, no. 1 (Summer 1980): 37-52.

——. "International Security and the Relevance of Arms Control." Daedalus 110, no. 1 (Winter 1980): 159-77.

Davis, Lynn Etheridge. Limited Nuclear Options: Deterrence and the New American Doctrine. Adelphi Papers no. 121. London: International Institute for Strategic Studies, Winter 1975/76.

Davis, Paul K. Land-Mobile ICBM's: Verification and Breakout. ACIS Working Paper no. 18. Los Angeles: Center for International and Strategic Affairs, UCLA, 1980.

Ermath, Fritz. "Contrasts in American and Soviet Strategic Thought." International Security 3, no. 2 (Fall 1978): 138-55.

Feld, Bernard T., and Kosta Tsipis. "Land-Based Intercontinental Ballistic Missiles." Scientific American 241, no. 5 (November 1979): 51-61.

Friedman, Norman. "The Soviet Mobilization Base." Air Force Magazine 62, no. 3 (March 1979): 65-71.

Garn, Jake. "SUM: Simplistic, Unworkable and Maladroit." Armed Forces Journal International, May 1980, pp. 55-56, 65.

Garwin, Richard L. "Launch Under Attack to Redress Minuteman Vulnerability?" International Security 4, no. 3 (Winter 1979/80): 117-39.

Gelb, Leslie H. "A Glass Half Full." Foreign Policy no. 36 (Fall 1979): 21-32.

Gelb, Leslie H., and Richard Ullman. "Keeping Cool at the Khyber Pass." Foreign Policy no. 34 (Spring 1979): 3-18.

Gray, Colin S. "The Arms Race Phenomenon." World Politics 24, no. 1 (October 1971): 39-79.

———. The Future of Land-Based Missile Forces. Adelphi Papers no. 140. London: International Institute for Strategic Studies, 1977.

———. "The Strategic Forces Triad: End of the Road?" Foreign Affairs 56, no. 4 (July 1978): 771-89.

———. "Nuclear Strategy: The Case for a Theory of Victory." International Security 4, no. 1 (Summer 1979): 54-87.

———. "Targeting Problems for Central War." Naval War College Review 33, no. 1 (January-February 1980): 3-21.

———. "Strategic Stability Reconsidered." Daedalus 109, no. 4 (Fall 1980): 135-54.

———. Strategy and the MX. Critical Issues series. Washington, D.C.: Heritage Foundation, 1980.

Gray, Colin S., and Keith Payne. "Victory is Possible." Foreign Policy no. 39 (Summer 1980): 14-27.

Griffith, William E. "Super-Power Relations After Afghanistan." Survival 22, no. 4 (July/August 1980): 146-51.

Hatfield, Mark. "SUM Strategy." Armed Forces Journal International, January 1980, p. 35.

———. "SUM: It Adds Up." Armed Forces Journal International, February 1980, p. 66.

Holloway, David. "Decision-Making in Soviet Defense Policies." In Prospects of Soviet Power in the 1980's, pt. 2, Adelphi Papers no.

152, pp. 24-31. London: International Institute for Strategic Studies, 1979.

——. "Military Power and Political Purpose in Soviet Policy." Daedalus 109, no. 4 (Fall 1980): 13-30.

Jervis, Robert. "Why Nuclear Superiority Doesn't Matter." Political Science Quarterly 94, no. 4 (Winter 1979/80): 617-33.

Kaiser, Robert, and Walter Pincus. "The Doomsday Debate: 'Shall We Attack America?'" Parameters 9, no. 4 (December 1969): 79-85.

Katz, Amron H. Verification and SALT: The State of the Art and the Art of the State. Washington, D.C.: Heritage Foundation, 1979.

Kincade, William H. "Will MX Backfire?" Foreign Policy no. 37 (Winter 1979-80): 43-58.

——. "A Farewell to Arms Control?" Arms Control Today 10, no. 1 (January 1980): 1-5.

Kissinger, Henry. "The Future of NATO." Washington Quarterly 2, no. 4 (Autumn 1979): 3-17.

Kugler, Jacek, and A. F. K. Organski, with Daniel Fox. "Deterrence and the Arms Race: The Impotence of Power." International Security 4, no. 4 (Spring 1980): 105-38.

Lambeth, Benjamin S. "The Political Potential of Soviet Equivalence." International Security 4, no. 2 (Fall 1979): 22-39.

Lee, William T. Soviet Defense Expenditures in an Era of SALT. Washington, D.C.: U.S. Strategic Institute, 1979.

——. "Debate over U.S. Strategic Forecasts: A Poor Record." Strategic Review 8, no. 3 (Summer 1980): 44-57.

Legvold, Robert. "Strategic 'Doctrine' and SALT: Soviet and American Views." Survival 21, no. 1 (January/February 1979): 8-13.

Lord, Carnes. "The ABM Question." Commentary 69, no. 5 (May 1980): 31-28.

Luttwak, Edward N. "Why Arms Control Has Failed." Commentary 65, no. 1 (January 1978): 19-28.

———. "SALT and the Meaning of Strategy." Washington Review 1, no. 2 (April 1978): 16-28.

———. "The American Style of Warfare and the Military Balance." Survival 21, no. 2 (March/April 1979): 57-60.

———. "After Afghanistan, What?" Commentary 69, no. 4 (April 1980): 40-49.

Meyer, Stephen M. "Verification and the ICBM Shell-Game." International Security 4, no. 2 (Fall 1979): 40-68.

Moll, Kendall K., and Gregory M. Luebbert. "Arms Race and Military Expenditure Models: A Review." Journal of Conflict Resolution 24, no. 1 (March 1980): 153-85.

"MX Still Zigzagging." Air Force Magazine 62, no. 2 (February 1979): 14.

"MX: The Missile We Don't Need." Defense Monitor 8, no. 9 (October 1979).

Nailor, Peter, and Jonathan Alford. The Future of Britain's Deterrent Force. Adelphi Papers no. 156. London: International Institute for Strategic Studies, 1980.

Nitze, Paul. "Assuring Strategic Stability in an Era of Detente." Foreign Affairs 54, no. 2 (January 1976): 207-32.

———. "Deterring Our Deterrent." Foreign Policy no. 25 (Winter 1976/77): 195-210.

Operations Research Society of America. "The Nature of Operations Research and the Treatment of Operations Research Quesions in the 1969 Safeguard Debate." Operations Research 19, no. 5 (September 1971): 1123-58; and 20, no. 1 (January-February 1972): 205-45.

Paine, Christopher E. "MX: The Public Works Project of the 1980's." Bulletin of the Atomic Scientists 36, no. 2 (February 1980): 12-16.

Pipes, Richard. "Why the Soviet Union Thinks It Could Fight and Win a Nuclear War." Commentary 64, no. 1 (July 1977): 21-34.

Ranger, Robin. "SALT II's Political Failure: The U.S. Senate Debate." RUSI Journal 125, no. 2 (June 1980): 49-56.

Rathjens, George W. "The Dynamics of the Arms Race." Scientific American 220, no. 4 (April 1969): 15-25.

Richelson, Jeffrey T. "Multiple Aim Point Basing: Vulnerability and Verification Problems." Journal of Conflict Resolution 23, no. 4 (December 1979): 613-28.

Rosen, Stephen. "Safeguarding Deterrence." Foreign Policy no. 35 (Summer 1979): 109-23.

Schneider, William, Jr. "Soviet Military Airlift: Key to Rapid Power Projection." Air Force Magazine 63, no. 3 (March 1980): 80-86.

Sienkiewicz, Stanley. "Observations on the Impact of Uncertainty in Strategic Analysis." World Politics 32, no. 1 (October 1979): 90-110.

Sigal, Leon V. "Rethinking the Unthinkable." Foreign Policy no. 34 (Spring 1979): 35-51.

Smith, Theresa C. "Arms Race Instability and War." Journal of Conflict Resolution 24, no. 2 (June 1980): 253-84.

Solzhenitsyn, Aleksandr. "Misconceptions About Russia Are a Threat to America." Foreign Affairs 58, no. 4 (Spring 1980): 797-834.

Steinbruner, John. "National Security and the Concept of Strategic Stability." Journal of Conflict Resolution 22, no. 3 (September 1978): 411-28.

Sullivan, David S. "The Legacy of SALT I: Soviet Deception and U.S. Retreat." Strategic Review 7, no. 1 (Winter 1979): 26-41.

————. Soviet SALT Deception. Boston, Va.: Coalition for Peace Through Strength, 1979.

Tsipis, Kosta. "The Accuracy of Strategic Missiles." Scientific American 223, no. 1 (July 1975): 14-23.

Wallop, Malcolm. "Opportunities and Imperatives of Ballistic Missile Defense." Strategic Review 6, no. 4 (Fall 1979): 13-21.

Wohlstetter, Albert J. "The Illusions of Distance." Foreign Affairs 46, no. 2 (January 1968): 242-55.

————. Legends of the Arms Race. Washington, D.C.: U.S. Strategic Institute, 1975.

POPULAR MAGAZINE/NEWSPAPER ARTICLES

Austin, Anthony. "Moscow Expert Says U.S. Errs on Soviet War Aims." New York Times, August 25, 1980, p. A2.

"Ballistic Missile Defense Tests Set." Aviation Week and Space Technology 112, no. 24 (June 16, 1980): 213-18.

Bassett, Edward W. "France to Modernize Nuclear Forces." Aviation Week and Space Technology 112, no. 24 (June 16, 1980): 265-69.

Biddle, Wayne. "The Silo Busters: Misguided Missiles: The MX Project." Harper's 259, no. 1555 (December 1979): 43-58.

Burt, Richard. "Likelihood of SALT's Demise Changes the Strategic Options." New York Times, March 23, 1980, p. E4.

————. "MX in Search of a Home and Mission." New York Times, March 31, 1980, p. A22.

————. "Soviet Nuclear Edge in Mid-80's Is Envisioned by U.S. Intelligence." New York Times, May 13, 1980, p. A12.

————. "Carter Said to Back a Plan for Limiting Any Nuclear War." New York Times, August 6, 1980, pp. A1, A6.

Clymer, Adam. "Behind Every Defense Policy There Lurks a Political Idea." New York Times, August 24, 1980, p. E4.

"Demonstration Planned for MX Defense System." Aviation Week and Space Technology 112, no. 24 (June 16, 1980): 220-21.

Evans, Rowland, and Robert Novak. "Unheeded Warnings About the ICBMs." Washington Post, January 12, 1979.

"Expert on Soviets Expects Trouble." New York Times, August 30, 1980, p. 5.

Gray, Colin S. "What the MX Would Do to Soviet Defense Planners." Letter to the editor. New York Times, April 16, 1980, p. A26.

Keller, Bill. "Attack of the Atomic Tidal Wave: Sighted S.U.M., Sank Same." Washington Monthly, May 1980, pp. 53-58.

"Kissinger's Critique." The Economist, February 3, 1979, pp. 17-22.

Lenorovitz, Jeffrey M. "Air Force Restudying MX Basing Plan." Aviation Week and Space Technology 110, no. 3 (January 15, 1979): 21.

Marsh, Alton K. "Pentagon Selects MX Grid Plan." Aviation Week and Space Technology 112, no. 19 (May 12, 1980): 15-16.

Mossberg, Walter S. "'Invisible' Plane Disclosure Looks Certain to Affect Soviet Relations, U.S. Politics." Wall Street Journal, August 25, 1980, p. 6.

"MX Associate Contractors Listed." Aviation Week and Space Technology 113, no. 2 (July 14, 1980): 70.

"Okay, on with MX." The Economist, May 10, 1980, p. 12.

Pond, Elizabeth. "Deterring Nuclear War: 2: The Soviet View." Christian Science Monitor, August 27, 1980, pp. 12-14.

Robinson, Clarence A., Jr. "Soviets Boost ICBM Accuracy." Aviation Week and Space Technology 108, no. 14 (April 3, 1978): 14-16.

———. "SALT 2 Approval Hinges on MX." Aviation Week and Space Technology 110, no. 20 (May 14, 1979): 14-16.

———. "Multipurpose Bomber Advances." Aviation Week and Space Technology 113, no. 5 (August 4, 1980): 16-18.

———. "Reagan Details Defense Boost." Aviation Week and Space Technology 113, no. 19 (November 10, 1980): 14-16.

"Serious Trouble Develops for Arms Treaty with Soviets." New York Times, September 28, 1979, p. A2.

Smith, Bruce A. "USAF Changes MX Missile Launch Mode." Aviation Week and Space Technology 112, no. 11 (March 17, 1980): 20–21.

———. "Opposition to MX Seen Growing in Utah." Aviation Week and Space Technology 112, no. 22 (June 2, 1980): 20–21.

———. "MX Basing Decisions Clear Way for Design Advances." Aviation Week and Space Technology 112, no. 24 (June 16, 1980): 132–34.

———. "Alternative MX Basing Poses Problems." Aviation Week and Space Technology 112, no. 26 (June 30, 1980): 22–23.

"Strategic Cuts Laid to Faulty Intelligence." Aviation Week and Space Technology 112, no. 8 (February 25, 1980): 19–20.

Wicker, Tom. "After the MX, What?" New York Times, March 25, 1980.

———. "Terror in Disguise." New York Times, August 24, 1980, p. E21.

Woolsey, R. James. "Getting the MX Moving." Washington Post, May 22, 1980, p. A17.

PUBLIC DOCUMENTS

Allen, General Lew, Jr. Letter to Melvin Price, chairman, Committee on Armed Services, U.S. House of Representatives, December 29, 1978.

———. Letter to John C. Stennis, chairman, Subcommittee on Defense, Committee on Appropriations, U.S. Senate, February 28, 1980.

Carter, Jimmy. "MX Missile System, Remarks on September 7th, 1979." Department of State Bulletin 79, no. 2032 (November 1979): 25–26.

CIA. Soviet Civil Defense. Washington, D.C.: CIA, 1978.

Congressional Budget Office. The MX Missile and Multiple Protective Structure Basing: Long-Term Budgetary Implications. Washington, D.C.: U.S. Government Printing Office, 1979.

Department of Defense Annual Report, Fiscal Year 1980. Washington, D.C.: U.S. Government Printing Office, 1979.

Department of Defense Annual Report, Fiscal Year 1981. Washington, D.C.: U.S. Government Printing Office, 1980.

Garwin, Richard L. Testimony to the Committee on Armed Services of the U.S. House of Representatives, February 7, 1979, pp. 29–39. (Mimeographed.)

Perry, William J. The Department of Defense Statement on the MX System and Ballistic Missile Defense. Before the Subcommittee on Research and Development of the Committee on Armed Services of the U.S. Senate, 96th Cong., 2nd sess. Washington, D.C.: Department of Defense, March 12, 1980.

Preliminary Draft Environmental Impact Statement for MX Deployment Area Selection and Land Withdrawal/Acquisition. Prepared for USAF, Ballistic Missile Office, Norton AFB, Calif. 9 vols. Santa Barbara, Calif.: HDR Sciences, 1980.

Rowen, Henry S. "Formulating Strategic Doctrine." In Commission on the Organization for the Conduct of Foreign Policy. Vol. 4, app. K, Adequacy of Current Organization: Defense and Arms Control, pp. 219–34. Washington, D.C.: U.S. Government Printing Office, 1975.

SALT One: Compliance/SALT Two: Verification. Selected Documents no. 7. Washington, D.C.: Department of State, 1978.

The Strategic Arms Limitation Talks. Special Report no. 46. Washington, D.C.: Department of State, 1978.

Undersecretary of Defense for Research and Engineering. ICBM Basing Options: A Summary of Major Studies to Define a Survivable Basing Concept for ICBM's. Washington, D.C.: Department of Defense, 1980.

U.S. Air Force. MX: Milestone II Draft Environmental Impact Statement. 5 vols. Norton AFB, San Bernardino, Calif.: Space and Missile Systems Organization, U.S. Air Force, 1978.

U.S. Department of State. SALT II Agreement. Selected Documents no. 12A, Department of State Publication 8984. Washington, D.C.: U.S. Government Printing Office, 1979.

U.S. House of Representatives, Committee on Armed Services, Panel
on the Strategic Arms Limitation Talks and the Comprehensive
Test Ban Treaty of the Intelligence and Military Application of Nuclear Energy Subcommittee. SALT II: An Interim Assessment.
Report. 95th Cong., 2nd sess. Washington, D.C.: U.S. Government Printing Office, 1978.

U.S. House of Representatives, Permanent Select Committee on In-
telligence, Subcommittee on Oversight. Soviet Strategic Forces,
Hearings. 96th Cong., 2nd sess. Washington, D.C.: U.S.
Government Printing Office, 1980.

U.S. Senate, Committee on Armed Services, Preparedness Investigating Subcommittee. Status of U.S. Strategic Power, Hearings.
Pt. 1. 90th Cong., 2nd sess. Washington, D.C.: U.S. Government Printing Office, 1969.

U.S. Senate, Committee on Foreign Relations. The SALT II Treaty,
Hearings. 96th Cong., 1st sess. Washington, D.C.: U.S. Government Printing Office, 1979.

U.S. Senate, Committee on Foreign Relations, Subcommittee on Arms
Control, International Law, and Organization. Briefing on Coun-
terforce Attacks, Hearings. 93rd Cong., 2nd sess. Washington,
D.C.: U.S. Government Printing Office, 1974.

U.S. Senate, Committee on Foreign Relations, Subcommittee on International Organization and Disarmament Affairs. Strategic and Foreign Policy Implications of ABM Systems, Hearings. Pt. 3. 91st
Cong., 1st sess. Washington, D.C.: U.S. Government Printing
Office, 1969.

Van Cleave, William R. "Solutions to ICBM Vulnerability." Prepared
Statement before U.S. Senate, Committee on Appropriations, Subcommittee on Military Construction, May 7, 1980.

Wolfowitz, Paul. "The Proposal to Launch on Warning." In U.S.
Senate, Committee on Armed Services, Authorization for Military
Procurement and Research and Development, Fiscal Year 1971,
and Reserve Strength, Hearings. Pt. 3, pp. 2278-82. 91st Cong.,
2nd sess. Washington, D.C.: U.S. Government Printing Office,
1970.

ADDITIONAL SOURCES

Brown, Harold. Speech at Naval War College, Newport, R.I., August 20, 1980.

Erickson, John. "The Soviet View of Nuclear War." Transcript of talk broadcast on BBC Radio 3, June 19, 1980.

Gray, Colin S. Arms Control and European Security: Some Basic Issues. HI-3157/2-P. Croton-on-Hudson, N.Y.: Hudson Institute, 1980.

——. Ballistic Missile Defense: A New Debate for a New Decade. HI-3160/3-P. Croton-on-Hudson, N.Y.: Hudson Institute, 1980.

Gray, Colin S., and Donald G. Brennan. Common Interests and Arms Control. HI-3218-P. Croton-on-Hudson, N.Y.: Hudson Institute, 1980.

Gray, Colin S., and Keith B. Payne. U.S. Strategic Posture and Deep Reductions in Force Levels Through SALT. HI-3195-RR. Croton-on-Hudson, N.Y.: Hudson Institute, 1980.

Halperin, Morton H. "Clever Briefers, Myopic Analysts, and Crazy Leaders." Unpublished paper prepared for the Aspen Arms Control Summer Study, June 1973.

Ikle, Fred. "The Prevention of Nuclear War in a World of Uncertainty." Speech given at the Joint MIT/Harvard Arms Control Seminar, Cambridge, Mass., February 20, 1974. (Mimeographed.)

Payne, Keith B. The BMD Debate: Ten Years After. HI-3040/2-P. Croton-on-Hudson, N.Y.: Hudson Institute, 1980.

Scott, William F., and Harriet F. Scott. The Soviet Control Structure. Final Report, SPC 575. Arlington, Va.: System Planning Corp., 1980.

Wohlstetter, Albert J., et al. Selection and Use of Strategic Air Bases. R-266. Santa Monica, Calif.: RAND, 1954.

INDEX

ABOUT THE AUTHOR

DR. COLIN GRAY is director of national security studies at Hudson Institute, Croton-on-Hudson, New York. Prior to joining Hudson Institute in 1976, he was an assistant director of the International Institute for Strategic Studies in London.

Dr. Gray has published extensively in professional journals in Europe and the United States. In 1981 he will publish Strategic Studies: A Critical Assessment and Strategic Studies and Public Policy: The American Experience. His previous books include The Soviet-American Arms Race (1976) and Canadian Defence Priorities (1972).

Dr. Gray received a B.A. (econ.) hons. in government from the University of Manchester in 1965, and a D. Phil. in international politics from Oxford University in 1970.